Tucholsky Wagner Zola Scott Sydow Schlegel
Turgenev Wallace Fonatne Freud
Twain Walther von der Vogelweide Fouqué Friedrich II. von Preußen
Weber Freiligrath Frey
Fechner Fichte Weiße Rose von Fallersleben Kant Ernst Richthofen Frommel
Engels Fielding Hölderlin
Fehrs Faber Flaubert Eichendorff Tacitus Dumas
Feuerbach Maximilian I. von Habsburg Fock Eliasberg Eliot Zweig Ebner Eschenbach
Ewald Vergil
Goethe London
Mendelssohn Balzac Shakespeare Elisabeth von Österreich
Lichtenberg Rathenau Dostojewski Ganghofer
Trackl Stevenson Tolstoi Hambruch Doyle Gjellerup
Mommsen Thoma Lenz Hanrieder Droste-Hülshoff
Dach Verne von Arnim Hägele Hauff Humboldt
Reuter
Karrillon Garschin Rousseau Hagen Hauptmann Gautier
Damaschke Defoe Descartes Hebbel Baudelaire
Wolfram von Eschenbach Schopenhauer Hegel Kussmaul Herder
Bronner Darwin Melville Dickens Grimm Jerome Rilke George
Campe Horváth Aristoteles Bebel Proust
Bismarck Vigny Gengenbach Barlach Voltaire Heine Federer Herodot
Storm Casanova Tersteegen Gilm Grillparzer Georgy
Chamberlain Lessing Langbein Gryphius
Brentano Lafontaine
Strachwitz Claudius Schiller Kralik Iffland Sokrates
Katharina II. von Rußland Bellamy Schilling
Gerstäcker Raabe Gibbon Tschechow
Löns Hesse Hoffmann Gogol Wilde Gleim Vulpius
Luther Heym Hofmannsthal Klee Hölty Morgenstern Goedicke
Roth Heyse Klopstock Kleist
Luxemburg Puschkin Homer Mörike
La Roche Horaz Musil
Machiavelli Kierkegaard Kraft Kraus
Navarra Aurel Musset
Nestroy Marie de France Lamprecht Kind Kirchhoff Hugo Moltke
Nietzsche Nansen Laotse Ipsen Liebknecht
Marx Ringelnatz
von Ossietzky Lassalle Gorki Klett Leibniz
May vom Stein Lawrence Irving
Petalozzi Platon Knigge
Sachs Poe Pückler Michelangelo Liebermann Kock Kafka
de Sade Praetorius Mistral Zetkin Korolenko

The Story of the Soil from the Basis of Absolute Science and Real Life,

Cyril G. (Cyril George) Hopkins

Imprint

This book is part of the TREDITION CLASSICS series.

Author: Cyril G. (Cyril George) Hopkins
Cover design: toepferschumann, Berlin (Germany)

Publisher: tredition GmbH, Hamburg (Germany)
ISBN: 978-3-8495-1315-3

www.tredition.com
www.tredition.de

The intention of the TREDITION CLASSICS series is to make world literature in the public domain available in printed format. Literary enthusiasts and organizations worldwide have scanned and digitally edited the original texts. tredition has subsequently formatted and redesigned the content into a modern reading layout. Therefore, we cannot guarantee the exact reproduction of the original format of a particular historic edition. Please also note that no modifications have been made to the spelling, therefore it may differ from the orthography used today.

TO MY WIFE

PREFACE

Truth is better than fiction; and this true story of the soil is written in co-operation with the Press of America and in competition with popular fiction.

The scenes described exist; the references given can all be found and verified; and the data quoted are exact, although some of the story dates antedate the scientific data.

As a rule the names employed are substitutes, but the general localities are as specified.

If the Story of the Soil should ever fall into the hands of any individual who suspects that he has contributed to its information, the author begs that he will accept as belonging to himself every gracious attribute and take it for granted that anything of opposite savor was due to autosuggestion.

CYRIL G. HOPKINS.

University of Illinois, Urbana.

CHAPTER I

THE OLD SOUTH

PERCY JOHNSTON stood waiting on the broad veranda of an old-style Southern home, on a bright November day in 1903. He had just come from Blue Mound Station, three miles away, with suit-case in hand.

"Would it be possible for me to secure room and board here for a few days?" he inquired of the elderly woman who answered his knock.

"Would it be possible?" she repeated, apparently asking herself the question, while she scanned the face of her visitor with kindly eyes that seemed to look beneath the surface.

"I beg your pardon, my name is Johnston,—Percy Johnston—" he said with some embarrassment and hesitation, realizing from her speech and manner that he was not addressing a servant.

"No pardon is needed for that name," she interrupted; "Johnston is a name we're mighty proud of here in the South."

"But I am from the West," he said.

"We're proud of the West, too; and you should feel right welcome here, for this is 'Westover,'" waving her hand toward the inroad fields surrounding the old mansion house. "I am Mrs. West, or at least I used to be. Perhaps the title better belongs to my son's wife at the present time; while I am mother, grandma, and great-grandmother.

"Yes, Sir, you will be very welcome to share our home for a few days if you wish; and we'll take you as a boarder. We used to entertain my husband's friends from Richmond,—and from Washington, too, before the sixties; but since then we have grown poor, and of late years we take some summer boarders. They have all returned to the city, however, the last of them having left only yesterday; so you can have as many rooms as you like.

"Adelaide!" she called.

A rugged girl of seventeen entered the hall from a rear room.

"This is my granddaughter, Adelaide, Mr. Johnston."

Percy looked into her eyes for an instant; then her lashes dropped. He remembered afterward that they were like her grandmother's, and he found himself repeating, "The eye is the window of the soul."

"My Dear, will you ask Wilkes to show Mr. Johnston to the southwest room, and to put a fire in the grate and warm water in the pitcher?"

"Thank you, that will not be necessary," said Percy. "I wish to see and learn as much as possible of the country hereabout, and particularly of the farm lands; and, if I may leave my suit-case to be sent to my room when convenient, I shall take a walk,—perhaps a long walk. When should I be back to supper."

"At six or half past. My son Charles has gone to Montplain, but he will be home for dinner. He knows the lands all about here and will be glad, I am sure, to give you any information possible."

With rapid strides Percy followed the private lane to the open fields of Westover.

"Is he a cowboy, Grandma?" asked Adelaide, in a tone which did not suggest a very high regard for cowboys. "Anyway," she continued, detecting a shade of disapproval in the grandmother's face, "he has a cowboy's hat, but he doesn't wear buckskin trousers or spurs."

Percy's hat was a relic of college life. Two years before he had completed the agricultural course at one of the state universities in the corn belt. Somewhat above the average in size, well proportioned, accustomed to the heaviest farm work, and trained in football at college, he was a sturdy young giant,—" strong as an ox and quick as lightning," in the exaggerated language of his football admirers

CHAPTER II

FORTY ACRES IN THE CORN BELT

PERCY JOHNSTON'S grandfather had gone west from "York State" and secured from the federal government a 160-acre "Claim" of the rich corn belt land. His father had received through inheritance only 40 acres of this; and, marrying his choice from the choir of the local Lutheran congregation, he had farmed his forty and an adjoining eighty acres, "rented on shares," for only three years, when he was taken with pneumonia from exposure and overwork, and died within a week.

Percy was scarcely a year old when his father was laid in the grave; but to the sorrowing mother he was all that life held dear. Existence seemed possible to her only because she could bestow upon him her double affection, and because the double duties which she took upon herself completely occupied her time.

She was not in immediate financial need, for her husband had been able to put some money in the bank during the last year, after having paid for his "outfit;" the forty-acre farm was free from debt, but under the law it must remain the joint property of mother and child for twenty years.

Wisely or unwisely she rejected every opportunity presented that would have given Percy a stepfather. As daughter and wife she had learned much of the art of agriculture, and, after some consultation with a neighbor who seemed to be successful, she made her own plans.

In her make up, sentiment was balanced with sense. Even as a young wife she had sometimes driven the mower or the self-binder to "help-out," and she had found pleasure and health in such hours of out-door life. "I can work and not overwork," she said to her friends; and in any case the crops seemed to grow better under the eye of the mistress.

Some years she employed a neighbor boy or girl, and always hired such other help as she needed. Prices were sometimes low and crops were not always good; and only widowed mothers can know the full story of her labor, love and sacrifice. With Percy's help he was sent to school and finally to the university, choosing for himself the agricultural college, much to the surprise and disappointment of his devoted mother.

"Why," she asked, "why should my son go to college to study agriculture? Have you not studied farming in the practical school of experience all your life? Surely we have done as much as could be done on our own little farm; and you have also had the benefit of the longer experience of our best farmers hereabout, and of the accumulated wisdom of our ancestors. Oh, I had hoped and truly believed that you would become interested in engineering, or in medicine, or may be in the law. I cannot understand why you should think of going to college to study farming. Surely you already know more than the college professors do about agriculture."

Percy's mother had too much good sense to have raised a spoiled boy. He had been taught to work and to think for himself. She loved her boy far better than her own life,—loved as only a widowed mother can who has risked her life for him, and who has given to him all her thought and all her energy from the best twenty years of her own life; but she had never let herself enjoy that kind of selfishness which prompts a mother to do for her child what he should be taught to do for himself. Despite his natural love of sport and the severe trials he had often brought to her patience and perseverance during his boyhood days, he had reached a development with the advance of youth that satisfied her high ideal. His love and appreciation and tender care for her repaid her every day, she told herself, for all the years of watching, working, waiting. Never before had he withstood her positive wish and final judgment.

And yet it was she who had told him that he alone must choose his life work and his college course in preparation for that work; but, after the years of toil, she had not dreamed that he would choose the farm life.

"My darling boy," she continued, "it leads to nothing. This little farm is poorer to-day than it was when your dear father and I came here to live and labor. To be sure, the lower field still grows as good or better crops than ever; but I can remember when that field was so wet and swampy that it could not be cultivated, and it was in the work of ditching and tiling that field," she sobbed, "that your father took the sickness that caused his death."

Tears were in Percy's eyes as he put his arm about his mother and wiped her tears away.

"But I must tell you what I know to be the truth," she went on quickly. "The older fields that your grandfather cultivated are less productive now than when he received them from our generous government. Indeed, it was your father's plan to continue to farm here only for a few years longer until he could save enough to enable him, with what we could have gotten from the sale of our own forty, to go farther west and purchase a large farm of virgin soil. He realized, my Son, that even that part of his father's farm that was first put under cultivation was becoming distinctly reduced in productiveness. He remembered, too, the stories often repeated by your grandfather of the run-down condition of the once exceedingly fertile soils of the Mohawk Valley and other parts of New York State.

"And you know, Percy, there were many Dutch farmers settled in New York. They were probably the best farmers among all who came to America from the Old World. I have heard your grandfather explain their use of crop rotation, and they understood well the value of clover and farm fertilizers. But with all of their skill and knowledge, the land grew poor, and now the very farm upon which Grandpa was born is not worth as much as the actual cost of the farm buildings. I hope you will consider all of this. The farm life is so unpromising for you, and there are such great opportunities for success in other lines. Still I feel that you must decide this question for yourself my Son, but tell me why you would choose the life and work of a farmer?"

CHAPTER III

LINCOLN S VIEW OF AGRICULTURE

PERCY had listened without interrupting, grieved at her disappointment, and open to any reasoning that might change his mind.

"Mother dearest," he said, "it was a year ago that you said I would have only till this fail to decide upon my college course and that it should be a special preparation for my life work. I have given much thought to it. You said that I should choose for myself, and I have not consulted much with others, but I have tried to consider the matter from different points of view.

"You know the Christmas present you gave me of the Lincoln books?"

"Yes, I know, and you have read them so much. I could not get you many books, but I knew there could be nothing better for my boy to read than the thoughts of that noble man. But, Percy dear, Lincoln was a lawyer, and he rose from the lowest walk in life to the highest position in the country, and with much less preparation than my own boy will have. Suppose he had remained a farmer! Surely no such success could ever have been reached. I am not so foolish as to have any such high hopes for you. Percy; but if you can only put yourself in the way of opportunity; and make such preparation as you can to fill with credit some position of responsibility that may be offered you! I had truly hoped that your study of Lincoln's life would influence yours. To me Lincoln was the noblest of all the noble men of our history, and I doubt not of all history, save Him who came to redeem the world."

Percy stepped to his little homemade bookcase and took a volume from the Lincoln set.

"May I read you some words of Lincoln?" he asked.

"Oh yes," she answered wonderingly.

"On September 30th, 1859," said Percy, "Lincoln gave an address at Milwaukee, before the State Agricultural Society of Wisconsin, and of all the addresses of Lincoln it seems to me that this is the greatest, because it deals with the greatest material problem of the United States. I think I have scarcely heard a public address in which the speaker has not dwelt upon the fact that the farmer must feed and clothe the world; and it seems to me that the missionaries always speak of the famines and starvation of so many people in India and other old countries. Do you remember the lecture by the medical missionary? Well, would it not he better to send agricultural missionaries to India and China to teach those people how to raise crops?

"I have read and reread this address more than any other in the Lincoln set. Let me read you some of the paragraphs I have marked.

"After making some introductory remarks about the value of agricultural fairs, Lincoln began his address as follows:

"'I presume I am not expected to employ the time assigned me in the mere flattery of the farmers as a class. My opinion of them is that, in proportion to numbers, they are neither better nor worse than other people. In the nature of things they are more numerous than any other class; and I believe there are really more attempts at flattering them than any other, the reason of which I cannot perceive, unless it be that they can cast more votes than any other. On reflection, I am not quite sure that there is not cause of suspicion against you in selecting me, in some sort a politician and in no sort a farmer, to address you.

"'But farmers being the most numerous class, it follows that their interest is the largest interest. It also follows that that interest is most worthy of all to be cherished and cultivated—that if there be inevitable conflict between that interest and any other, that other should yield.

"'Again, I suppose that it is not expected of me to impart to you much specific information on agriculture. You have no reason to believe, and do not believe, that I possess it; if that were what you seek in this address, any one of your own number or class would be more able to furnish it. You, perhaps, do expect me to give some general interest to the occasion, and to make some general suggestions on practical matters. I shall attempt nothing more. And in such suggestions by me, quite likely very little will be new to you, and a large part of the rest will be possibly already known to be erroneous.

"'My first suggestion is an inquiry as to the effect of greater thoroughness in all the departments of agriculture than now prevails in the Northwest—perhaps I might say in America. To speak entirely within bounds, it is known that fifty bushels of wheat, or one hundred bushels of Indian corn, can be produced from an acre.'"

Percy paused: "You know, Mother, that our corn has averaged some less than fifty bushels per acre for the last five years, and, as you say, the lower field has been much better than the old land, and I think you are quite right in your belief that as an average the land is growing poorer, although we cultivate better than we used to do, and our seed corn is of the best variety and saved with much care. But let me read further:

"'Less than a year ago I saw it stated that a man, by extraordinary care and labor, had produced of wheat what was equal to two hundred bushels from an acre. But take fifty of wheat, and one hundred of corn, to be the possibility, and compare it with the actual crops of the country. Many years ago I saw it stated, in a patent office report, that eighteen bushels was the average crop throughout the United States; and this year an intelligent farmer of Illinois assured me that he did not believe the land harvested in that State this season had yielded more than an average of eight bushels to the acre; much was cut, and then abandoned as not worth threshing, and much was abandoned as not worth cutting.'"

"I know it is true," said the mother, "that wheat was once very much grown in Central and Northern Illinois, but 1859 must have been an unusually poor year, for it was grown for twenty years after that, although it finally failed so completely that its cultivation has been practically abandoned in those sections for nearly twenty years. However, the chinch bugs were a very important factor in discouraging wheat growing and the land has been very good for corn, especially since the tile-drainage was put in; but on the whole is it not as I told you?"

"But note these statements," said Percy, turning again to the book:

"'It is true that heretofore we have had better crops with no better cultivation, but I believe that it is also true that the soil has never been pushed up to one-half of its capacity.

"'What would be the effect upon the farming interest to push the soil up to something near its full capacity?'"

"But what can he mean," said the mother. "How can anyone do better than we have done? We change our crops, and sow clover with the oats, and return as much as we can to the land. But let me hear further what Lincoln said:"

"Yes, Mother, this is what he said:

"'Unquestionably it will take more labor to produce fifty bushels of wheat from an acre than it will to produce ten bushels from the same acre; but will it take more labor to produce fifty bushels from one acre than from five? Unquestionably thorough cultivation will require more labor to the acre; but will it require more to the bushel? If it should require just as much to the bushel, there are some probable, and several certain, advantages in favor of the thorough practice. It is probable it would develop those unknown causes which of late years have cut down our crops below their former average. It is almost certain, I think, that by deeper plowing, analysis of the soils, experiments with manures and varieties of seeds, observance of seasons, and the like, these causes would be discovered and remedied. It is certain that thorough cultivation would spare half, or more than half, the cost of land, simply because the same produce would be got from half, or from less than half, the quantity of land. This proposition is self-evident, and can be made no plainer by repetitions or illustrations. The cost of land is a great item, even in new countries, and it constantly grows greater and greater, in comparison with other items, as the country grows older.'"

Percy paused and said: "If I understand correctly these words of Lincoln, the land need not become poor. But I do not know why land becomes poor. I do not know what the soil contains, nor do I know what corn is made of. We plow the ground and plant the seed and cultivate and harvest the crop, but I do not know what the corn crop, or any crop, takes from the soil. I want to learn how to analyze the soil and crop and to find out, if possible, why soils become poor, in order, as Lincoln suggests, that the cause may be discovered and remedied."

"It may be that the college professors could teach you in that way," said the mother, "but you know the farm life is so full of work and so empty of mental culture."

"I used to think so too," said Percy, "but I fear we have worked too much with our hands and too little with our minds; that we have done much work in blindness as to the actual causes that control our crop yields; and that we have not found the mental culture that may be found in the farm life. Let me read again. These are Lincolns words:

"'No other human occupation opens so wide a field for the profitable and agreeable combination of labor with cultivated thought, as agriculture. I know nothing so pleasant to the mind as the discovery of anything that is at once new and valuable—nothing that so lightens and sweetens toil as the hopeful pursuit of such discovery. And how vast and how varied a field is agriculture for such discovery! The mind, already trained to thought in the country school, or higher school, cannot fail to find there an exhaustless source of enjoyment. Every blade of grass is a study; and to produce two where there was but one is both a profit and a pleasure. And not grass alone. but soils, seeds, and seasons—hedges, ditches, and fences—draining, droughts, and irrigation—plowing, hoeing, and harrowing—reaping, mowing, and threshing—saving crops, pests of crops, diseases of crops, and what will prevent or cure them—implements, utensils, and machines, their relative merits, and how to improve them—hogs, horses, and cattle—sheep, goats and poultry—trees, shrubs, fruits, plants, and flowers—the thousand things of which these are specimens—each a world of study within itself.

"'In all this book learning is available. A capacity and taste for reading gives access to whatever has already been discovered by others. It is the key, or one of the keys, to the already solved problems. And not only so; it gives a relish and facility for successfully pursuing the unsolved ones. The rudiments of science are available, and highly available. Some knowledge of botany assists in dealing with the vegetable world—with all growing crops. Chemistry assists in the analysis of soils, selection and application of manures, and in numerous other ways. The mechanical branches of natural philosophy are ready help in almost everything, but especially in reference to implements and machinery.

"'The thought recurs that education—cultivated thought—can best be combined with agricultural labor, on the principle of thorough work; that careless, half-performed, slovenly work makes no place for such combination; and thorough work, again, renders sufficient the smallest quantity of ground to each man; and this, again, conforms to what must occur in a world less inclined to wars and more devoted to the arts of peace than heretofore. Population must increase rapidly, more rapidly than in former times, and ere long the most valuable of all arts will be the art of deriving a comfortable subsistence from the smallest area of soil. No community whose every member possesses this art, can ever be the victim of oppression in any of its forms. Such community will be alike independent of crowned kings, money kings, and land kings.'"

CHAPTER IV

LIFE'S CHOICE

PERCY read these words as though they were his own; and perhaps we may say they were his own, for, as Emerson says: "Thought is the property of him who can entertain it."

The mother listened, first with wonder; then with deepened interest, which changed to admiration for the language and for her son, who seemed to be filled with the spirit which had led Lincoln to see the problems and the possibilities of the farm life in a light that was wholly new.

"Surely those are noble thoughts," she said, "from a noble and wise man. I shall only hope that you will find some opportunity to make the best possible of your life. We have such a small farm, and the land hereabout is all so high in price that to enlarge the farm seems almost hopeless. In part because of this difficulty it had seemed to me that greater opportunities might be open for you in other lines. Don't you feel that you will be greatly handicapped in the beginning?"

"Perhaps," said Percy, "in some ways; but not in other ways. We hear on every hand that this is an age of specialists, that the most successful man cannot take time to prepare himself well for many different lines of work; that he must make the best possible preparation in some one line for which he may have special talent or special interest; and then endeavor to go farther in that line than any one has gone before. When I first wrote to the State University I asked how long a time would likely be required for me to complete all the subjects that are taught there, and the registrar replied that, if I could carry heavy work every year, I might hope to take all the courses now offered in about seventy years. In considering this point of preparation for future work, it has seemed to me that if I leave the farm life and devote myself to law or to engineering, I must in large measure sacrifice about ten years of valuable experience in practical agriculture. I have learned enough about farming so that I can manage almost as well as the neighbors; and without this knowledge, gathered, as you say, in the school of experience, I can see that serious mistakes would often be made.

"You know that Doctor Miller bought the Bronson farm two years ago. Well, he has been giving some directions himself concerning its management. He has had no experience in farming, and last year, after he had the new barn built, he directed his men to put the sheaf oats in the barn so they would be safe from the weather. He did not understand that oats must stand in the shock for two or three weeks to become thoroughly "cured" before they can safely be even stacked out of doors; and the result was that his entire oat crop rotted in the barn.

"People who have lived always in the city sometimes express the most amusing opinions of farm conditions so well understood even by a ten-year-old country boy. I recently overheard two traveling men remarking about the differences which they could plainly observe between the corn crops in different fields as they rode past in the train.

"'Some fields have twice as good corn as other adjoining fields,' one remarked. 'How do you account for the difference,' asked the other. 'oh, I suppose the one farmer was too stingy of his seed,' was the reply.

"I am convinced that there are hundreds or perhaps thousands of valuable facts that have been acquired through experience and observation by the average farm boy of eighteen or twenty years that would be of little or no value to him in most other occupations; and in this respect I should be handicapped if I leave the farm life and begin wholly at the bottom in some other profession. Perhaps agriculture is not a profession, but I think it should be if the highest success is to be attained."

"I surely hope you will be successful, Percy, and your reasoning sounds all right; but other occupations seem to lead to greater wealth than farming."

"I very much doubt," replied Percy, "if there is any other occupation that is so uniformly successful as farming, in the truest sense. It provides constant employment, a good living, and a comfortable home for nearly all who engage in it; and as a rule they have made no such preparation as is required for most other lines of work.

"But there is still another side to the farm life, Mother dear, or to any life for that matter. Your own life has taught me that to work for the love of others is a motive which directs the noblest lives. If agricultural missionaries are needed in India, they are also needed in parts of our own country where farm lands that were once productive are now greatly depleted and in some cases even abandoned for farming; and. if the older lands of the corn belt are already showing a decrease in productive power, we need the missionary even here. If I can learn how to make land richer and richer and lead others to follow such a system, I should find much satisfaction in the effort."

CHAPTER V

WORN OUT FARMS

"WELL, you found some mighty poor land, I reckon," was the greeting Percy received from Grandma West as he returned from his walk over Westover and some neighboring farms.

"I found some land that produces very poor crops," he replied, "but I don't know yet whether I should say that the land is poor."

"Well, I know it's about as poor as poor can be; but it was not always poor, I can tell you. When I was a girl, if this farm did not produce five or six thousands bushels of wheat, we thought it a poor crop; but now, if we get five or six hundred bushels, we think we are doing pretty well. My husband's father paid sixty-eight dollars an acre for some of this land, and it was worth more than that a few years later and, mind you, in those days wheat was worth less and niggers a mighty sight more than they are nowadays; but, somehow, the land has just grown poor. We don't know how. We have worked hard, and we have kept as much stock as we could, but we could never produce enough fertilizer on the farm to go very far on a thousand acres.

"Yes, Sir, we have just about a thousand acres here and we still own it,—and with no mortgage on it, I'm mighty glad to say. But, laws, the land is poor, and you can get all the land you want about here for ten dollars an acre. There comes Charles, now. He can tell you all about this country for more than twenty miles, I reckon.

"Wilkes!" A negro servant answered the call, and took the horse as Charles West stopped at the side gate.

"Wilkes was born here in slave times, nigh sixty years ago," she continued. "He is three years older than my son Charles. He has remained with us ever since the war, except for a few months when he went away one time just to see for sure that he was free and _could _go. But he came back mighty homesick and he'll want to stay here till he dies, I reckon.

"Charles, this is Mr. Johnston, Percy Johnston, as he says; but he thinks he is no kin of General Joe or Albert Sidney. He's been looking at the land hereabout, but I don't think he'll want any of it after seeing the kind of crops we raise."

With this introduction, the mother disappeared within the house, and
Charles took her seat on the vine-covered veranda.

"I feel that I owe an apology to you, Sir," said Percy, "for presenting myself here with bag and baggage, and asking to share the hospitality of your home, with no previous arrangements having been made; but by chance I met your friend, Doctor Goddard, on the train, and, in answer to my inquiry as to whom I could go to for correct information concerning the history and present condition and value of farm lands in this section of the country, he advised me to stop off at Blue Mound Station and consult with you. Had I known that you were to be in Montplain to-day, of course I should have gone directly there. Your mother very graciously consented to receive me as a belated summer boarder, a kindness which I greatly appreciate, I assure you.

"My mother and I have a small farm in Illinois,—so small that it would be lost in such an estate as Westover, but the price of land is very high in the West at the present time; and I am really considering the question of selling our little forty-acre farm and purchasing two or three hundred acres in the East or South. My thought is that I might secure a farm that was once good land, but that has been run down to such an extent that it can be bought for perhaps ten or twenty dollars an acre. I should want the land to be nearly level so that it would not be difficult to prevent damage from surface washing. I should prefer, of course, to purchase where there is a good road and not more than five miles from a railway station.

"If I secure such a farm, it would be my purpose to restore its fertility. If possible I should want to make the land at least as productive as it ever was, even in its virgin state."

"Well, Sir," said Mr. West, "if you could accomplish your purpose and ultimately show a balance on the right side of the ledger, it would be a work of very great value to this country. There will be no difficulty in securing such land as you want with location and price to suit you; but I think that you should know in advance that older men than you have purchased farms hereabout with very similar intentions, but with the ultimate result that they have lost more, financially, than we who are native to the soil; for, while we were once well-to-do and are now poor, we still own our land, impoverished as it is. However, the farm still furnishes us a comfortable living, supplemented, to be sure, with some income from other sources.

"I am very willing to give as much information as I can regarding our lands and the agricultural conditions and common practices, although I fear that this knowledge will discourage you from making any investments in our worn-out farms. If you still decide to make the trial, I surely hope you will be successful, for we need such an object lesson above all else.

"I assume that you will wish to locate near a town of considerable size, in order that you can haul manures from town, and perhaps some feed also; and have a good market for your milk and other products."

"No, Sir," said Percy, "I should prefer not to engage in dairying, and I do not wish to make use of fertilizer made from my neighbors' crops. We have some object lessons of that kind in my own state; and I have no doubt that some can be found in this state who feed all they produce on their own land and perhaps even larger amounts of feed purchased from their neighbors, or hauled from town, and who, in addition to using all of the farm fertilizer thus produced, haul considerable amounts of such materials from the livery stables in town. With much hard work, with a good market for the products of the dairy and truck garden, and with business skill in purchasing feed from their neighbors when prices are low, such men succeed as individuals; but do they furnish an object lesson which could be followed by the general farmer?"

"I had not looked at the matter from that point of view," said Mr. West, "but it is plain to see that on the whole there can be only a small percentage of such farmers; and in reality they are a detriment to their neighbors who permit their own hay and grain to be hauled off from their farms; but certainly these are the methods followed by our most successful farmers, and these are they who live on the fat of the land."

"Are they farmers or are they manufacturers?" asked Percy. "It seems to me that, in large measure, their business is to manufacture a finished product from the raw materials produced upon other farms, either in the immediate neighborhood or in the newer regions of the West. As you know, much of our surplus produce from the farms of the corn belt is shipped into the eastern and southern states, there to be used as food for man and beast, not only in the cities, but also to a considerable extent in the country. Instead of living on the fat of the land, such manufacturers live in the country at the expense of special city customers who may have fat jobs and are able to pay fancy prices for country produce made by the impoverishment of many farms. In most cases, if such a 'successful farmer' were compelled to pay average prices for what he buys and allowed to receive only average prices for what he sells, his fat would have plenty of lean streaks."

CHAPTER VI

THE MUSICALE

DINNER was served at the family table, with Mr. West at the head and his mother at the foot.

"The eye is the window of the soul," thought Percy, as he met the glance of Adelaide sitting opposite. Certain he was that he had never before looked into such alluring eyes.

Adelaide was neither a girl nor a woman and yet at times she was both. With the other children she was a child that still loved to romp and play with the rest, free as a bird. Her mother, a sweet-faced woman, some years her husband's junior, made sisters of all her daughters, the more naturally perhaps, because the grandmother was still so active and so interested in all phases of homemaking that she seemed mother to them all. Adelaide's two older sisters were married and her brother Charles, also older than herself, by three years, was a senior in college. Adelaide had just finished her course in the Academy where the long service of a maiden aunt as a teacher had secured certain appreciated privileges, without which it is doubtful if both Charles and Adelaide could have been sent away to school at the same time. A boy of fourteen and the eight-year old baby brother with two sisters between comprised the younger members of the family.

Miss Bowman, the teacher of the district school, also occupied a place at the table. The evening meal was disposed of without delay, for there was something of greater importance to follow. A musicale in the near-by country church had been in preparation and Percy heartily accepted an invitation to accompany the family to the evening's entertainment. Or rather he accompanied Mr. and Mrs. West and the grandmother, for all the children had walked the distance before the carriage arrived.

Without having specialized in music, nevertheless Percy had improved the frequent opportunities he had had, especially while at the university, and he had learned to appreciate quality in the musical world. Consequently he was not a little surprised and greatly pleased to sit and listen to a class of music that he had never before heard rendered in country places; but, as he listened for Adelaide's singing in chorus, duet, and solo, he found himself wondering whether the eye or the voice more clearly revealed the soul.

"It seemed like the old times," said the grandmother, with something like a sigh, as she took her place in the carriage. "If our land was only like it used to be! but it's become so mighty poor our children can't have many advantages these days. The Harcourt's and Staunton's whom you met are descendants of ancestors once well known in this state."

"It seems to me that the land need not have grown poor," said Percy. "If the land was once productive, its fertility ought to be maintained by the return of the essential materials removed in crops or destroyed by cultivation. Surely land need not become poor; but of course I know too little about this land to suggest at the present time what method could best be adopted for its improvement."

"We can tell you what the best method is," she quickly replied. "Just put on plenty of ordinary farm fertilizer, but, laws, we don't have enough to cover fifty acres a year."

For a time each seemed lost in thought, or listening to the husband and wife who sat in the front seat quietly talking of the evening's performances. Percy recognized some of the names they mentioned as belonging to persons to whom he had been presented at the church. It gradually dawned upon him that he had spent the evening with the aristocracy of the Blue Mound neighborhood. Culture, refinement, and poverty were the chief characteristics of the people who had been assembled.

"It need not have been," he repeated to himself; "surely, it need not have been, "and then he wondered if these were not much sadder words than the oft repeated "it might have been."

"May I ask where your people came from, Mrs. West?" he questioned.

"Where we came from?" she repeated, "I don't quite understand."

"Excuse me," said Percy, "but in the West it is so common to ask people where they are from. You know the West is settled with people from all sections of the East, and many from Europe and from Canada, and I thought your ancestors may have moved here from some other state, as from Pennsylvania for example, where my mother's people once lived."

"Let me advise you, Young Man," said the grandmother briskly, and in a tone that reminded Percy of the twinkle he had at times noticed in her eyes when she seemed young again—"Let me advise you never to ask a Virginian if he was born in Pennsylvania. That's more than most Virginians can stand. Once a Virginian, always a Virginian,—both now, hereafter, and hitherto. It's mighty hard to find a Virginian who came from anywhere except from the royal blood of England; although some may condescend to acknowledge kinship to the Scottish royalty."

The grandmother's voice was raised to a pitch which commanded the attention of the other members in the carriage and a hearty laugh followed her jovial wit, to the full relief of Percy's temporary embarrassment.

"Well," she continued, "to answer your question: my husband and my children are direct descendants of Colonel Charles West, a brother of Lord Delaware, who was Sir Thomas West, whose ancestry goes back to Henry the Second, of England, and to David the First, of Scotland; and my granddaughter is the great-granddaughter of Patrick Henry. So now you know where _we _came from," and she laughed again like a girl. "Yes," she added, "we have a family tree six feet from branch to branch, but it is stored in a back room where I am sure it is covered with cobwebs, for we have no time to live with the past when the summer boarders are here."

As the carriage stopped at the side gate, the children's voices could be heard in the rear; for Mr. West had been living over again his younger days with his sweet-faced wife, and the farm team had taken its own time.

CHAPTER VII

A BIT OF HISTORY

"NOW, I shall be at home to-day and glad to assist you in any way possible," announced Mr. West at the breakfast table.

"That is very kind of you," Percy replied. "I want especially to learn some of the things you know about the soils of Westover. Can you show me the best land and the poorest land on the estate?"

"I think I can." said Mr. West. "We have some land that has not grown a crop in fifty years, and we have other land that still produces a very fair crop if properly rotated."

"And what rotation do you practice?"

"Well, the system we have finally settled into and have followed for many years is to plow up the run-out pasture land and plant to corn. The second year we usually raise a crop of wheat or oats and seed down to clover and timothy. We then try to cut hay from the land for two years, and afterward we use the field for pasture for six or eight years, or until finally it produces only weeds and foul grass. Then we cover it with farm manure, so far as we can, and again plow the land for corn. Wheat and cattle are the principal products sold from the farm."

"In this way," said Percy, "you grow one crop of corn on the same field about once in ten or twelve years."

"Yes, about that, and also one, or sometimes two, crops of small grain. We usually have about seventy-five acres of corn, nearly a hundred acres of small grain, and we cut hay from somewhat more than hundred acres, thus leaving perhaps five hundred acres of pasture land, besides about two hundred acres of timber land which has not been cultivated for many years."

"Was the timber land that we see about here formerly cultivated?" asked Percy.

"Oh, yes, nearly all of it was under cultivation when I was a boy, although some had been allowed to go back to timber even before I was born. On our own farm we have some timber land that, so far as I have been able to learn, was never under cultivation; and the character of the trees is different on that land. There you will find original pine, but on the worn-out land the 'old-field' pine are found. They are practically worthless, while the original pine makes very valuable lumber.

"With our system of rotation we keep about all of our farm under control; but the smaller farms were necessarily cropped more continuously to support the family, and they became so unproductive that many of them have been completely abandoned for agricultural purposes; and even some of the large plantations were poorly managed, one part having been cropped continuously until too poor to pay for cropping, while the remainder was allowed to grow up in scrub brush and 'old-field' pine; and, of course, the expense of clearing such land is about as much as the net value of the crops that could be grown until it again becomes too poor for cropping."

"Then the recleared lands are not as productive as when they were first cleared from the virgin forest?"

"Oh, by no means. In the virgin state these lands grew bountiful crops almost continuously for a hundred years or more. Virginia was famed at home and abroad for her virgin fertility. Great crops of corn, wheat, and tobacco were grown. Tobacco was a valuable export crop, and there were many Virginians whose mothers came to America with passage paid for in tobacco. History records, you may remember, that it was the custom for a time to permit a young man to pay into a general store house a hundred pounds of tobacco,—and this was later increased to one hundred fifty pounds,—to be used in payment of passage for young women who were thus enabled to come to America; and there was a very distinct understanding that only those who had come forth with the tobacco were eligible as suitors for the hand of any 'imported' maiden. As a matter of fact some such arrangement as this was almost a necessity," said Mr. West, as he noted Adelaide's almost incredulous look. "Among the first settlers in Virginia, young men greatly predominated; and in the main the people in the home country were themselves in poverty. Under the hereditary laws of England the father's estate and title became the possession of his eldest son; and in large measure the other children of the family were thrown absolutely upon their own resources, so that many, even with royal blood in their veins, were very glad to embrace any opportunity offered to seek a new home in this land of virgin richness.

"Of course," he continued, smilingly and in direct answer to Adelaide's inquiring look, "those young women were in no sense bound to accept the attention or the offer of any man; but naturally most of them did become the wives of those who were able to offer them a husband's love and a home with more of life's comforts perhaps than they had ever known before. They were at perfect liberty, however, to remain in the enjoyment of single blessedness if they chose, and I doubt not," he added, with a twinkle in his eyes, "that some of them had no other choice."

CHAPTER VIII

WESTOVER

WITH an auger in his hand, by means of which a hole could be quickly bored into the soil to a depth of three or four feet, Percy joined Mr. West for the tramp over the plantation.

In general the estate called Westover consists of undulating upland. A small stream crosses one corner of the farm bordered by some twenty acres of bottom land which is subject to frequent overflow, and used only for permanent pasture. Several draws or small valleys are tributary to the stream valley, thus furnishing excellent surface drainage for the entire farm. In some places the sides of these valleys are quite sloping and subject to moderate erosion when not protected by vegetation. Above and between these slopes the upland is nearly level. As they came upon one of these level areas, grown up with small forest trees, Mr. West stopped and said:

"Now right here is probably as poor a piece of land as there is on the farm. This land will positively not grow a crop worth harvesting unless it is well fertilized."

"If we were in the Illinois corn belt," replied Percy, "I should expect to find the land in this position to be the most productive on the farm. Our level uplands are now valued at from one hundred fifty to two hundred dollars an acre. A farm of one hundred eighty acres, five miles from town, sold for two hundred and fourteen dollars an acre a few days before I started east."

"Well," said Mr. West, "this may have been good land once, but if so it was before my time. Of course most of our uplands here have been cropped for upwards of two hundred years; and about all that has ever been done to keep up the fertility of the soil has been to rotate the crops. To be sure, the farm manure has always been used as far as it would go, but the supply is really very small compared to the need for it."

"Do you think that the proper rotation of crops would maintain the fertility of the soil?" asked Percy.

"No, I have tried too many rotations to think that, but I suppose it is a help in that direction, don't you?"

"I would say that crop rotation may help to maintain the supply of some important constituents of a fertile soil, but it will certainly hasten the depletion of some other equally essential constituents."

"Well, that's a new idea to me. I may not quite grasp your meaning; but first tell me about these tests you are making."

When they stopped on the area of poor land as designated by Mr. West, Percy had turned his auger into the earth and drawn out a sample of moist soil, which he molded into the form of a ball. He broke this in two, inserted a piece of blue paper, and pressed it firmly together. He then laid the ball of soil aside, secured another sample with the auger, and formed it into a cake with a hollow in the upper surface. He took from his pocket a slender box or tube of light wood, removed the screw cap, and drew out a glass-stoppered bottle.

"This bottle contains hydrochloric acid," said Percy. "It is often incorrectly called 'muriatic acid.' It consists of two elements, hydrogen and chlorin, from which its name is derived. But you are perhaps already familiar with the chemical elements."

"Well, I heard lectures at William and Mary for four years, and they included some chemistry as it was then taught; but they certainly did not include the application of chemistry to agriculture, and I am greatly interested to know the meaning of these tests you are making here on our own farm under my own eyes. You may take it for granted that I know absolutely nothing of such use of chemistry as you are evidently turning to some practical value."

"Any other farmer can make these tests as well as I can," said Percy. "This bottle of acid cost me fifteen cents and it can be duplicated for the same price at almost any drug store. The acid is very concentrated, in fact about as strong as can easily be produced, but it need not be especially pure. Some care should be taken not to get it on the clothing or on the fingers, although it is not at all dangerous to handle, but it tends to burn the fingers unless soon removed, either by washing with water or by rubbing it off with the moist soil."

"I use this acid to test the soil for the presence or absence of limestone. Ordinary limestone consists of calcium carbonate. Here, again the chemical name alone is sufficient to indicate the elements that compose this compound. It is only necessary to keep in mind the fact that the ending *-ate* on the common chemical names signifies the presence of oxygen Thus calcium carbon_ate_ is composed of the three primary elements, calcium, carbon and oxygen.

"Of course the chemical element is the simplest form of matter. An element is a primary substance which cannot be divided into two or more substances All known matter consists of about eighty of these primary elements; and, as a matter of fact, most of these are of rare occurence—many of them much more rare than the element gold.

"About ninety-eight per cent. of the soil consists of eight elements united in various compounds or combinations; and only ten elements are essential for the growth and full development of corn or other plants. If any one of these ten elements is lacking, it is impossible to produce a kernel of corn, a grain of wheat, or a leaf of clover; and in the main the supply is under the farmer's own control. But we can discuss this matter more fully later. Let us see what we have here."

Percy poured a few drops of the hydrochloric acid into the hollow of the cake of soil.

"What should it do?" asked Mr. West.

"If the soil contains any limestone, the acid should produce foaming, or cffervescence," replied Percy; "but it is very evident that this soil contains no limestone. You see the hydrochloric acid has power to decompose calcium carbonate with the formation of carbonic acid and calcium chlorid, a kind of salt that is used to make a brine that won't freeze in the artificial ice plants. The carbonic acid, if produced at once decomposes into water and carbon dioxid. Now, the liberated carbon dioxid is a gas and the rapid generation or evolution of this gas constitutes the bubbling or foaming we are looking for; but since there is no appearance of foaming we know that this soil contains no limestone."

"Then you have already found that those three elements,—calcium, carbon, and oxygen, you called them, I think—you find that those elements are all lacking in this soil."

"No, this test does not prove that," said Percy. "It only proves that they are not present as limestone. Calcium may be present in other compounds, especially in silicates, which are the most abundant compounds in the soil and in the earth's crust; and, as indicated by the ending _-ate, _oxygen is contained in calcium silicate as well as in calcium carbonate."

"I see; the subject is much more complicated than I thought."

"Somewhat, perhaps," Percy replied; "but yet it is quite simple and very easily understood, if we only keep in mind a few well established facts. Certainly the essential science of soil fertility is much less complicated than many of the political questions of the day, such as the gold standard or free-silver basis, the tariff issues, and reciprocity advantages, regarding which most farmers are fairly well informed,—at least to such an extent that they can argue these questions for hours."

"I think you are quite right in that," said Mr. West. "Of course, it is important that every citizen entitled to the privilege of voting in a democracy like ours should be able to exercise his franchise intelligently; but the citizen who is responsible for the management of farm lands ought surely to be at least as well informed concerning the principles which underlie the maintenance of soil fertility; provided, of course, that such knowledge is within his reach; and from what you say I am beginning to believe that such is the case. At any rate this simple test seems to show conclusively that this soil contains no limestone, and it is common knowledge that limestone soils are good soils."

Percy took up the ball of soil containing the slip of blue paper, broke it in two again, and it was seen that the paper had changed in color from blue to red

"There's a change, for certain," said Mr. West, "that has some meaning to you I suppose."

"This is litmus paper," said Percy. "It is prepared by moistening specially prepared paper with a solution of a coloring matter called litmus, and the paper is then dried. This coloring matter has the property of turning blue in the presence of alkali and red in the presence of acid. The blue paper is prepared with a trace of alkali, and the red paper with a trace of acid. If more than a trace were present the litmus paper would not be sufficiently sensitive for the test.

"This little bottle containing two dozen slips of paper cost me five cents, and it can be obtained at most drug stores.

"Alkali and acid are exactly opposite terms, like hot and cold. The one neutralizes the other. This test with litmus paper is a test for soil acidity, and the fact that the moisture of the soil has turned the litmus from blue to red shows that this soil is acid, or sour. The soil moisture contained enough acid to neutralize the trace of alkali contained in the blue paper and to change the paper to a distinctly light red color; and the fact that the paper remains red even after drying, shows that the soil contains fixed acids or acid salts, and not merely carbonic acid, which if present would completely volatilize as the paper dries.

"Now, these two tests are in harmony. The one shows the absence of limestone, and the other shows the presence of acidity, and consequently the need of limestone to correct or neutralize the acidity, for limestone itself is an alkali."

"But limestone soils are not alkali soils, are they?" asked Mr. West.

"Not in the sense of containing injurious alkali, like sodium carbonate, the compound which is found in the 'black alkali' lands of the arid regions of the far West; but chemically considered limestone is truly an alkali; and, as such, it has power to neutralze this soil acidity."

"Is the acidity harmful to the crops?"

"It is not particularly harmful to the common crops of the grass family, such as wheat, corn, oats, and timothy; but some of the most valuable crops for soil improvement will not thrive on acid soils. This is especially true of clover and alfalfa."

"That is certainly correct for clover so far as this kind of soil is concerned," said Mr. West. "Clover never amounts to much on this kind of land, except where heavily fertilized. When fertilized it usually grows well. Does the farm fertilizer neutralize the acid?"

"Only to a small extent. It is true that farm manures contain very appreciable amounts of lime and some other alkaline, or basic, substances, but in addition to this, and perhaps of greater importance, is the fact that such fertilizer has power to feed the clover crop as well as other crops. In other words it furnishes the essential materials of which these crops are made. In addition to this the decaying organic matter has power to liberate some plant food from the soil which would not otherwise be made available although to that extent the farm manure serves as a soil stimulant, this action tending not toward soil enrichment but toward the further depletion of the store of fertility still remaining in the soil."

"This seems a complicated problem," said Mr. West, "but may I now show you some of our more productive land?"

"As soon as I collect a sample of this," replied Percy, and to Mr. West's surprise he proceeded to bore about twenty holes in the space of two or three acres. The borings were taken to a depth of about seven inches, and after being thoroughly mixed together an average sample of the lot was placed in a small bag bearing a number which Percy recorded in his note book together with a description of the land.

"I wish to have an analysis made of this sample," remarked Percy, as they resumed their walk.

"But I thought you had analyzed this soil," was the reply.

"Oh, I only tested for limestone and acidity," explained Percy. "I wish to have exact determinations made of the nitrogen and phosphorus, and perhaps of the potassium, magnesium, and calcium. All of these are absolutely essential for the growth of every agricultural plant; and any one of them may be deficient in the soil, although" the last three are not so likely to be as the other two."

"How long will it take to make this analysis?" was asked.

"About a week or ten days. Perhaps I shall collect two or three other samples and send them all together to an analytical chemist. It is the only way to secure positive knowledge in advance as to what these soils contain. In other words, by this means we can take an absolute invoice of the stock of fertility in the soil, just as truly as the merchant can take an invoice of the stock of goods carried on his shelves."

"So far as we are concerned, this would not be an invoice in advance," remarked Mr. West, with a shade of sadness in his voice. "If we knew the contents of the crops that have been sold from this farm during the two centuries past, we would have a fairly good invoice, I fear, of what the virgin soil contained; but can you compare the invoice of the soil with that of the merchant's goods?"

"Quite fairly so," Percy replied. "The plant food content of the plowed soil of an acre of normal land means nearly, if not quite, as much in the making of definite plans for a system of permanent agriculture, as the merchant's invoice means in the future plans of his business.

"It should not be assumed that the analysis of the soil will give information the application of which will always assure an abundant crop the following season. In comparison, it may also be said, however, that the merchant's invoice of January the first may have no relation to the sales from his store on January the second. Now, the year with the farmer is as a day with the merchant. The farmer harvests his crop but once a year; while the merchant plants and harvests every day, or at least every week. But I would say that the invoice of the soil is worth as much to the farmer for the next year as the merchant's invoice is to him for the next month.

"It should be remembered, however, that both must look forward, and plans must be made by the merchant for several months, and by the farmer for several years. Your twelve-year rotation is a very good example of the kind of future planning the successful farmer must do. On the other hand, some of your neighbors, who have not practiced some such system of rotation now have 'old-field' pine on land long since abandoned, and soil too poor to cultivate on land long cropped continuously."

"This is a kind soil," remarked Mr. West, as he paused on a gently undulating part of the field.

"That is a new use of the word to me," said Percy. "Just what do you mean by a 'kind' soil?"

"Well, if we apply manure here it will show in the crop for many years. It is easy to build this soil up with manure; but, of course, we have too little to treat it right."

"The soil is almost neutral," said Percy, testing with litmus and acid. "Does clover grow on this soil?"

"Very little, except where we put manure."

Another composite sample of the soil was collected, and they walked on.

"Now, here," said Mr. West, "is about the most productive upland on the farm."

"Is that possible?" asked Percy, the question being directed more to himself than to his host.

"That is according to my observation for about fifty years," he replied. "Where we spread the farm fertilizer over this old pasture land and plow it under for corn, we often harvest a crop of eight barrels to the acre, while the average of the field will not be more than five barrels.—A barrel of corn with us is five bushels."

They had stopped on one of the steepest slopes in the field.

"These hillsides would be considered the poorest land on the farm if we were in the corn belt," said Percy, "but I think I understand the difference. Your level uplands when once depleted remain depleted, because the soil that was plowed two hundred years ago is the same soil that is plowed to-day; but these slopes lose surface soil by erosion at least as rapidly as the mineral plant food is removed by cropping; and to that extent they afford the conditions for a permanent system of agriculture of low grade, unless, of course, the erosion is more rapid than the disintegration of the underlying bed rock, which I note is showing in some outcrops in the gullies.

"I want some samples here," he continued, and at once proceeded to collect a composite sample of the surface soil and another of the sub-soil.

"In the main this soil is slightly acid," said Percy, after several tests, with the hydrochloric acid and the litmus paper; "although occasionally there are traces of limestone present. The mass of soil seems to be faintly acid, but here and there are little pieces of limestone which still produce some localized benefit, and probably prevent the development of more marked acidity throughout the soil mass.

"If I can get to an express office this afternoon," he continued, "I shall be glad to forward these four composite samples to an analyst."

"If you wouldn't mind riding to Montplain with Adelaide when she goes for her music lesson this afternoon, it would be very convenient," said Mr. West.

"With your daughter's permission that would suit me very well," he replied. "I shall be glad to spend one or two days more in this vicinity, and then I wish to visit other sections for a week or two, after which I would be glad to stop here again on my return trip and probably I shall have the report of the chemist concerning these samples."

CHAPTER IX

THE BLACK PERIL

AS Percy stepped out of the house in the early afternoon upon the announcement from Wilkes that "De ca'age is ready," he noted that the "ca'age" was the two-seated family carriage and that Adelaide had already taken her place in the front seat, as driver, with her music roll and another bundle tucked in by her side. Her glance at Percy and at the rear seat was also sufficient to indicate his place.

"This does not seem right to me, Miss West," said Percy. "Unless you prefer to drive I shall be very glad to do so and let you occupy this more comfortable seat."

"No thank you," she replied, in a tone that left no room for argument. "I often drive our guests to and from the station, and I much prefer this seat."

The rear seat was roomy and low, so that Percy could scarcely see the road ahead even by sitting on the opposite side from the driver.

Aside from an occasional commonplace remark both the driver and the passenger were allowed to use the time for meditation.

While Adelaide was already an experienced horsewoman, she was rarely permitted to drive the colts to the village, although she enjoyed riding the more spirited horses, or driving with her brother in the "buck board."

A mile from the village the road wound through a wooded valley, and then climbed the opposite slope, passing the railway station a quarter of a mile from town and the "depot hotel" near by. Here

Percy left the carriage with the bags of soil, it being arranged that he would be waiting at the hotel when Adelaide returned from the village.

Adelaide's "hour" was from four to five, and being the last pupil for the day, the teacher was not prompt to close.

"I did not realize the days were becoming so short," said Miss Konster as she opened the door. "I'm sorry you have so far to drive."

"Oh, I don't mind," said Adelaide, "I know the way home well enough. You see I have the double carriage, for I brought a guest to the depot as usual, although he is to return with me, and is probably very tired of waiting at the 'depot hotel.'"

It was nearly dark as Percy took his place in the rear seat, Adelaide having again declined to yield her position as driver, and now she had more packages nearly filling the seat beside her.

The team leisurely took the homeward way and nothing more was said except an occasional word of encouragement to the horses. They passed the lowest point in the valley and began to ascend the gentle slope, when the carriage suddenly stopped, and Adelaide uttered a muffled scream. "Come, Honey, said a masculine voice."

As Percy half rose to his feet, he saw that a negro had grasped Adelaide in an effort to drag her from the carriage. A blow from Percy staggered the brute and he released his hold of Adelaide, but, as he saw Percy jump from the carriage on the opposite side, he paused.

"De's a man heah. Knock him, Geo'ge," he yelled, as he turned to again grapple with Adelaide

"Coward," cried Adelaide, as she saw Percy jump from the carriage and dart up the road. Facing this black brute, she was standing alone now with one hand on the back of the seat. As the negro sprang at her the second time he uttered a scream like the cry of a beast and fell sprawling on his face. Almost at the same moment his companion was fairly lifted from his feet and came down headlong beside the carriage.

"Look out for the horses," called Percy, as he drove the heels of his heavy shoes into the moaning mass on the ground.

"Lie there, you brute," he cried, "don't you dare to move."

"I have the lines," said Adelaide hoarsely, "but can't I do something more?"

"No. they're both down," he answered. "Wait a minute."

He found himself between the negroes lying with their faces to the ground. Instantly he grasped each by the wrist and with an inward twist he brought forth cried for mercy. It was a trick he had learned in college, that, by drawing the arm behind the back and twisting, a boy could control a strong man.

"Can't I help you?" Adelaide called again, and Percy saw that she was out of the carriage and standing near.

"Will the horses stand?" he asked.

"Oh, yes, they're quiet now."

"Then take the tie rope and tie their feet together. Use the slip knot just as you do for the hitching post," he directed. "If they dare to move I can wrench their arms out in this position. Right there at the ankles. Tie them tight and as closely together as you can. Wrap it twice around if it's long enough."

Adelaide tied one end of the rope around the ankle of one negro and wrapped the other end around the ankle of the other, drawing their feet together and fastening the ends of the rope with a double hitch, which she knew well how to make.

Percy gave the rope a kick to tighten it.

"Now get onto your feet and I'll march you to town," he ordered, adding pressure to the twist upon their wrists and drawing them back upon their knees Thus assisted, they struggled to their feet.

"I am afraid you will have to drive home alone, Miss West," began Percy, when Adelaide interrupted with:

"No, no, if you are going back to town, I will follow you. I can easily turn the team and I will keep close behind."

Thus tied together, Percy almost ran his prisoners toward the village, still holding each firmly by the wrist. As they reached the "depot hotel," he called for assistance, and several men quickly appeared.

Percy made a brief report of the attack as they moved on to the town house, where the villians were placed in shackles and left in charge of the marshall.

"Will you drive, please, Mr. Johnston?" asked Adelaide as he stepped to the carriage; for Adelaide had followed almost to the door of the jail house.

"Yes, please," he replied, taking the seat beside her.

"I hope you will pardon my calling you a coward, I felt so desperate, and it seemed to me for the moment that you were leaving me." Adelaide's voice still had an excited tremor to it.

"I heard you say 'coward,'" said Percy, "but I didn't realize that you referred to me. I saw the two brutes almost at the same time, the one who attacked you and the other on the same side near the horses' heads. I struck the one as best I could from my position, and as he yelled and the horses reared, I ran up the slope ahead of the team and came down at the other brute with a blow in the neck, but I was surprised to find them both sprawling on the ground; and under the street lights I saw that one of them had an eye frightfully jammed. I am sure I struck neither of them in the eye."

Adelaide made no reply, but she knew now that the piercing, beastly cry from the negro reaching for her was brought forth because the heel of her shoe had entered the socket of the brute's eye.

"You're mighty nigh too late for supper, said grandma West, as they stopped at the side gate. Adelaide hurried to her father who took her in his arms as he saw how she trembled.

"My child!" he said.

Yes, child she was as she relaxed from the tension of the last hour and related the experience of the evening.

"I cannot express our gratitude to you, Sir," said Mr. West: "I am glad you landed the devils in jail."

"I am only thankful I was there when it happened," replied Percy. "I am sure no man could have done less. I have promised to return to town in the morning to serve as legal witness in the case. I hope your daughter need not be called upon for that."

"Probably that will not be necessary," Mr. West replied.

CHAPTER X

THE SLAVE AND THE FREEDMAN

THE others had retired but Percy and his host continued their conversation far into the night.

"There are almost as great variations among the negroes as among white people," Mr. West was saying. "To a man like Wilkes who was born and raised here on the farm, I would entrust the protection of my wife and children as readily as to any white man. He has been educated, so to speak, to a sense of duty and honor; and negroes of his class have almost never been known to violate a trust. Of course there are bad niggers, but as a rule such negroes have grown up under conditions that would develop the evil in any race of men.

"During the Secession it was the most common thing for the men to go to war and leave their defenseless women and children wholly in the care of their slaves; and, even though the federal soldiers were fighting to free the slaves and their masters to keep them in slavery, rarely did a negro fail to remain faithful to his trust. They hid from the northern soldiers the horses and mules, cotton and corn, clothing and provisions, and all sorts of valuables; and in most cases were ready to suffer themselves before they would reveal the hidden property. To be sure there were masters who abused their slaves, and some of these were naturally ready to desert at the first opportunity; but in the main the slave owner was more kind to his human property than the considerate soldier was to his horse, and the negro as a race is appreciative of kindness."

"I suppose the depreciation in soil fertility and crop yields dates largely from the freeing of the slaves does it not?" asked Percy.

"Well, that was one factor, but not the most potential factor. Much land in the south had been abandoned agriculturally long before the war, and much land in New York and New England has been abandoned since the war. The freeing of the negroes produced much less effect in the economic conditions of the south than many have supposed. The great injury to the South from the war was due to the war itself and not to the freeing of slaves. In the main it cost no more to hire the negro after the war than it cost to feed and clothe him before; and the humane slave owner had little difficulty in getting plenty of negro help after the war. Very commonly his own slaves remained with him and were treated as servants, not particularly differently than they had been treated as slaves. Of course there were some brutal slave holders, just as there are brutal horse owners, and such men suffered very much from the loss of slave labor.

"The southern people have no regrets for the freeing of the slaves. Probably it was the best thing that ever happened to us; and the South would have less regret for the war itself, except that our recovery from it was greatly delayed by the reconstruction policy which was followed after the war. The immediate enfranchisement of the negro, especially in those sections where this resulted in placing all the power of the local government in the hands of the negro, was a worse blow to the South than the war itself.

"It is believed that this would not have been done if Lincoln had lived. Lincoln was always the President of all the people of the United States, and his death was a far greater loss to the South than to the North. To place the power to govern the intelligent white of the South absolutely in the hands of their former ignorant slaves was undoubtedly the most abominable political blunder recorded in history; and even this was intensified by the unprincipled white-skinned vultures who came among us to fatten upon our dead or dying conditions. Those years of so-called reconstruction, constitute the blackest page in the history of modern civilization."

"I quite agree with you," said Percy, "and so far as I know them the soldiers of the northern armies also agree with you. Several of my own relatives fought to free the negro slave; but none of them fought to enslave their white brothers of the South by putting them absolutely under negro government. And yet there is one possible justification for that abominable reconstruction policy. It may have averted a subsequent war which might have lasted not for four years, but for forty years. Even if this be true, perhaps there is no credit in the policy for any man who helped to enforce it, but you will grant that there were two important results from those bitter years of reconstruction:

"First, the negro learned with certainty at once and forever that he was a free man.

"Second, he at once acquired a degree of independence effectually preventing the development of a situation throughout the South, in which the negro, though nominally free, would have remained virtually a slave, a situation which, if once established, might have required a subsequent war of many years for its complete eradication. Even under the conditions which have prevailed, there have been isolated instances of peonage in the southern states since the war; and if the education and gradual enfranchisement of the negro had been left wholly in the hands of their former masters, from the immediate close of the war, I can conceive of conditions under which slavery would essentially have been continued."

"Such a possibility is, of course, conceivable," said Mr. West, "and we must all admit that there were some slave holders who would have taken advantage of any such opportunity; but had Lincoln lived the terms made would probably have been such that the South would have felt in honor bound to enforce them. Probably the enfranchisement would have been based upon some sort of qualification such as the southern states have very generally adopted in subsequent years; but the idea of social equality of slave and master was so repulsive to the white people of the South that it could not be tolerated under any sort of government."

"This question of social equality," remarked Percy, "has probably been the cause of more misunderstanding between the North and the South than all other questions relating to the negro problem. I have rarely, if ever, talked with a southern man who did not have it firmly fixed in his mind that the common idea of the northern people is that the negro race should be made the social equal of the white race. This I have heard from southern lecturers; I have read it in southern newspapers; and I have found it in books written by southern authors; but, Mr. West, I have never yet heard that idea advanced by a man or woman of the North.

"Of course there have been visionary theorists or 'cranks' in all ages, and there must have been some basis for this almost universal erroneous opinion in the South that the people of the North advocated social equality or social intercourse between the white and colored races; and yet nothing could be farther from the truth. In all my life in the North, I think I have never seen a colored person dining with a white man. This does not prove that there are no such occurrences, but it certainly shows that they are extremely rare. On the other hand, in traveling through the South I have seen a white woman bring her colored maid or nurse, to the dining car and sit at the same table with herself and husband. Of course there is no suggestion of social equality or social intercourse in this, but there is a much closer relationship than is common or would be allowed in the North."

"That may be true," said Mr. West, "and there was in slave times a very intimate relationship between the negro nurses and the white children of the South. Some of our people are ready to take offense at the suggestion that we talk negro dialect, and perhaps we would all prefer to say that the negroes have learned to talk as we talk; but the truth is that the negroes were brought to America chiefly as adults; and, as is usually the case when adult people learn a new language, they modified ours because their own African language did not contain all of the sounds of the English tongue. Similarly we hear and recognize the other nationalities when they learn to speak English. Thus we have the Irish brogue, the German brogue, and the French brogue, or dialect.

"The negro children learned to speak the dialect as spoken by their own parents; and as a very general rule the white children learned to talk as their negro nurses talked. So far as there is a southern dialect it is due to the modification of our language by the negro."

"You have mentioned several things," said Percy, "that are much to the credit of the negro who has had a fair chance to be trained along right lines; and I think the modficaton of our language which his presence has brought about in the South is not without some credit. It is generally agreed that the most pleasing English we hear is that of the Southern orator.

"Referring to social conditions, the most marked difference which I have noticed between the North and South, and really, it seems to me, the only difference of importance, is that the South has separate schools for white and colored, whereas in the North the school is not looked upon as a social institution.

"As a rule no more objection is raised to white and colored children sitting on separate seats in the same school room than to their sitting on separate seats in the same street car. The school is regarded as a place for work, where each has his own work to do, much the same as in the shop or factory where both white and colored are employed. The expense of the single school system is, of course, much less than where separate schools are maintained; and perhaps an equally important point is that in the single system the same moral standards are held up by the teachers for both white and colored children."

"That point is worthy of consideration," said Mr. West. "It is very certain that a class of negroes has grown up in these more recent years that was practically unknown in slave times when white men were more largely responsible for their moral training. The vile wretches who made the attack this evening probably never received any moral training. It is conceivable that the moral influence of the white children over the negroes in the same school might exert a lasting benefit, even aside from the influence of the teacher; and the relationship of the school room could not be any real disadvantage to the white child. But this could only be brought about where white teachers were employed. Some such arrangement would doubtless have been made had the mind of Lincoln directed the general policy of reconstruction; but it is doubtful now if the negro teacher will ever be wholly replaced, although time has wrought greater changes in political lines since the black years of the reconstruction."

"Yes," said Percy, "and those changes which have been brought about in the South have the full sympathy and approval of the great majority of the Northern people. Indeed, it is extremely doubtful if the North will be able to completely banish such a source of vice and corruption as the open saloon until limitation is placed upon the franchise by an educational qualification."

CHAPTER XI

JUDGMENT IS COME

THE goddess of sleep seemed to have deserted Westover. Adelaide lay in her mother's arms, either awake and restless or in fitful sleep from which she frequently awoke with a muffled scream or a physical contortion. Once, as she nestled closer, her mother heard her murmur: "You must pardon me."

Percy, from the southwest room, was sure he heard horses feet at the side gate. The murmur of low voices reached his ear, and then he recognized that horsemen were riding away.

The house was astir at early dawn; and as soon as breakfast was over Mr. West had the colts hitched to the "buckboard" and he drove with Percy to Montplain.

"I think your testimony will not be needed this morning," said Mr. West, "but it may be needed later, and it is well that you should report to the officers at any rate, since you promised to be there this morning."

Percy pointed out the place where the attack had been made, and he looked for a stump of a small tree or for any other object upon which the negro could have fallen with such force as to mash his eye; but he saw nothing.

As soon as they reached the village, Mr. West drove directly to the town house; and there two black bodies were seen hanging from the limb of an old tree in the courthouse yard. Percy noted that his companion showed no sign of surprise; and, after the first shock of his complete realization of the work of the night, he looked calmly upon the scene. They had stopped almost under the tree.

"Are these the brutes who made the attack and whom you captured and delivered to the officer?" asked Mr. West.

"They are," he replied.

"In your opinion have they received justice?"

"Yes, Sir," Percy replied, "but I fear without due process of law."

"Let me tell you, Sir, there is no law on the statutes under which justice could be meted out to these devils for the nameless crime which ends in death by murder or by suicide of the helpless victim, a crime which these wretches committed only in their black hearts—thanks to you, Sir."

As he spoke, the town marshall approached followed by the negro pastor of the local church and a few of his followers. Silently they lowered the bodies to the ground, placed them upon improvised stretchers, and carried them to the potters field outside the village, where rough coffins and graves were ready to receive them.

As Mr. West and Percy returned to Westover they discussed the lands which in the main were lying abandoned on either side of the road.

"Here," said Mr. West, as he paused on the brow of a sloping hillside, "was as near to Westover as the Union army came. The position of the breastworks may still be seen. The Southern army lay across the valley yonder. These two trees are sprouts from an old stump of a tree that was shot away. About seventeen hundred confederate dead were buried in trenches in the valley, but they were later removed. The federal dead were carried away as the Union army retreated. We never learned their number. For three days Westover was made headquarters of the confederate officers, and my mother worked day and night to prepare food for them."

They stopped at Westover for a few moments, Percy remaining in the "buckboard" while Mr. West reported to his family what they had seen in Montplain.

"Our report," said Mr. West, "hideous and horrible as it is, will help to restore the child to calm and quiet. To speak frankly, Sir, occurrences of this sort, sometimes with the worst results, are sufficiently frequent in the South so that we constantly feel the added weight or burden whenever the sister, wife, or daughter is left without adequate protection."

The remaining hours of the morning were devoted to a drive over the country surrounding Westover; and Mr. West consented to Adelaide's request that she be allowed to drive Percy to the station at Montplain, where he was to take the afternoon train for Richmond. She chose the "buckboard" but insisted upon driving.

They talked of their school and college days, of the books they had read, of anything in fact except of the experiences of the past twenty-four hours. Even when they entered the valley no shadow crossed Adelaide's face; but as they neared the station her voice changed, and as Percy looked into her winsome, frankly upturned face, she said:

"Have I truly been pardoned for my cruel words last evening? I am sure you were as manly and noble as any man could have been."

"And I am sure you were the bravest little woman I have ever known," replied Percy, "and I admire you the more for calling me a coward when you thought I was running away; so there is nothing to pardon I am sure."

She gave him her hand as a child at parting, but he thought as he looked into her eyes that he saw the soul of a woman.

CHAPTER XII

THE RESTORATION

PERCY carried with him a most interesting and attractive circular of information concerning the rapid restoration of the farm lands of the South. It also stated that further information could be secured from a certain real estate agent in Richmond, who was found to be still in his office when Percy arrived in the city late in the afternoon.

The agent was delighted to receive a call from the Western man, and assured him that he would gladly show him several plantations not far from the city which could be purchased at very reasonable prices. Indeed he could have his choice of these old southern homesteads for the very low price of forty dollars an acre. A map of an adjoining county showed the exact location of several such farms, some of which were of great historical interest. At what time in the morning could he be ready to be shown one of these rare bargains?

"What treatment do these lands require to restore their productiveness?" asked Percy.

"No treatment at all, Sir, except the adoption of your western methods of farming and your system of crop rotation. I tell you the results are marvelous when western farmers get hold of these famous old plantations. Just good farming and a change of crops, that's all they need."

"Does clover grow well?" asked Percy. "We grow that a good deal in the
West."

"Oh, yes, clover will grow very well, indeed, but cowpeas is a much better crop than clover. Our best farmers prefer the cowpea; and after a crop of cowpeas, you can raise large crops of any kind."

"Of course you know of those who have been successful in restoring some of these old farms," Percy suggested.

"Oh, yes, Sir, many of them, and they are making money hand over fist, and their lands are increasing in value, and no doubt will continue to increase just as your western lands have done. Yes, Sir, the greatest opportunity for investment in land is right here and now, and these old plantations are being snapped up very rapidly."

"I shall be glad to know of some of these successful farmers who are using the improved methods. Will you name one, just as an example, and tell me about what he has done to restore his land?"

"Well," said the agent, "There's T. O. Thornton, for example. Mr. Thornton bought an old plantation of a thousand acres only six years ago at a cost of six dollars an acre. He has been growing cowpeas in rotation with other crops; and, as I say, he is making money hand over fist. A few months ago he refused to consider fifty dollars an acre for his land, but still there are some of these old plantations left that can be bought for forty dollars, because the people don't really know what they are worth. However, our lands are all much higher than they were a few years ago."

"Where does Mr. Thornton live?" asked Percy.

"Oh, he lives at Blairville, nearly a hundred miles from Richmond. Yes, he lives on his farm near Blairville. I tell you he's making good all right, but I don't know of any land for sale in that section."

"I think I will go out to Blairville to see Mr. Thornton's farm," said Percy. "Do you know when the trains run?"

"Well, I'm sorry to say that the train service is very poor to Blairvile. There is only one train a day that reaches Blairville in daylight, and that leaves Richmond very early in the morning."

"That is all right," said Percy, "it will probably get me there in time so that I shall be sure to find Mr. Thornton at home. I thank you very much, Sir. Perhaps I shall be able to see you again when I return from Blairville."

"When you return from Blairville is about the most uncertain thing in the world. As I said, the train service is mighty poor to Blairville, and it's still poorer, you'll find, when you want to leave Blairville. Why, a traveling man told me he had been on the road for fifteen years, and he swore he had spent seven of 'em at Blairville waiting for trains. Better take my advice and look over some of the fine old plantations right here in the next county and then you can take all the rest of the month if you wish getting in and out of Blairville."

About eight o'clock the following morning Percy might have been seen walking along the railroad which ran through Mr. Thornton's farm about two miles from Blairvile. He saw a well beaten path which led from the railroad to a nearby cottage and a knock brought to the door a negro woman followed by several children.

"Can you tell me where Mr. Thornton's farm is?" he inquired.

"Yes, Suh," she replied. "This is Mistah Tho'nton's place, right heah, Suh. Leastways, it was his place; but we done bought twenty acahs of it heah, wheah we live, 'cept tain all paid fo' yit. Mistah Tho'nton lives in the big house over theah 'bout half a mile."

"May I ask what you have to pay for land here?"

"Oh, we have to pay ten dollahs an acah, cause we can't pay cash. My ol' man he wo'ks on the railroad section and we just pay Mistah Tho'nton foh dollahs every month. My chil'n wo'k in the ga'den and tend that acah patch o' co'n."

"Do you fertilize the corn?"

"Yes, Suh. We can't grow nothin' heah without fe'tilizah. We got two hundred pounds fo' three dollahs last spring and planted it with the co'n."

As Percy turned in at Mr. Thornton's gate he saw a white man and two negroes working at the barn. "Pardon me, but is this Mr. Thornton?" asked Percy as he approached.

"That is my name."

"Well, my name is Johnston. I am especially interested in learning all I can about the farm lands in this section and the best methods of farming. I live in Illinois, and have thought some of selling our little farm out there and buying a larger one here in the East where the land is much cheaper than with us. A real estate agent in Richmond has told me something of the progress you are making in the improvement of your large farm. I hope you will not let me interfere with your work, Sir."

"Oh, this work is not much. I've had a little lumber sawed at a mill which is running just now over beyond my farm, and I am trying to put a shed up here over part of the barn yard so we can save more of the manure. I shall be very glad to give you any information I can either about my own farming or about the farm lands in this section."

"You have about a thousand acres in your farm I was told."

"Yes, we still have some over nine hundred acres in the place, but we are farming only about two hundred acres, including the meadow and pasture land. The other seven hundred acres are not fenced, and, as you will see, the land is mostly grown up to scrub trees."

"Your corn appears to be a very good crop. About how many acres of corn do you have this year?"

"I have only fourteen acres. That is all I could cover with manure, and it is hardly worth trying to raise corn without manure."

"Do you use any commercial fertilizer?"

"Well, I've been using some bone meal. I've no use for the ordinary complete commercial fertilizer. It sometimes helps a little for one year; but it seems to leave the land poorer than ever. Bone meal lasts longer and doesn't seem to hurt the land. I see from the agricultural papers that some of the experiment stations report good results from the use of fine-ground raw rock phosphate; but they advise using it in connection with organic matter, such as manure or clover plowed under. I am planning to get some and mix it with the manure here under this shed. Do you use commercial fertilizers in Illinois?"

"Not to speak of, but some of our farmers are beginning to use the raw phosphate. Our experiment station has found that our most extensive soil types are not rich in phosphorus, and has republished for our benefit the reports from the Maryland and Ohio experiment stations showing that the fine-ground natural rock phosphate appears to be the most economical form to be used and that it is likely to prove much more profitable in the long run, although it may not give very marked results the first year or two. May I ask what products you sell from your farm, Mr. Thornton?"

"I sell cream. I have a special trade in Richmond, and I ship my cream direct to the city. I also sell a few hogs and some wheat. I usually put wheat after corn, and have fourteen acres of wheat seeded between the corn shocks over there. Sometimes I don't get the wheat seeded, and then I put the land in cowpeas. I usually raise about twenty-five acres of cowpeas, and the rest of the cleared land I use for meadow and pasture. I usually sow timothy after cowpeas, and I like to break up as much old pasture land for corn as I can put manure on."

"I was told that you had been offered fifty dollars an acre for your farm, Mr. Thornton, but that you would not consider the offer."

Mr. Thornton laughed heartily at this remark.

"That must have come from the Richmond land agent," he said. "Someone else was telling me that story a short time ago. The fact is one of those real estate agents was out here last spring and he asked me if I would consider an offer of fifty dollars an acre for our land. I told him that I didn't think that I would as long as any one who wishes to buy can get all the land he wants in this section for five or ten dollars an acre. That's as near as I came to having an offer of fifty dollars an acre for this land. The land adjoining me on the south is is for sale, and I am sure you could buy that farm of about seven hundred acres for four dollars an acre after they get the timber off. Some of the land has not been cropped for a hundred years, I guess; and there are a few trees on it that are big enough for light saw-stuff. A man has bought the timber that is worth cutting, and he is running a saw over there now; but he'll get out all that's good for anything in a few months."

"May I ask how long you have been farming here, Mr. Thornton?"

"Twelve years on this farm," he replied. "You see this estate was left to my wife and her sister who still lives with us. We were married twelve years ago and I have been working ever since to make a living for us on this old worn-out farm. Of course I have made some little improvements about the barns, but we've sold a little land too. The railroad company wanted about an acre down where that little stream crosses, for a water supply, and I got twelve hundred dollars for that."

"Now, I've already taken too much of your time," said Percy. "I thank you for your kindness in giving me so much information. If there is no objection I shall be glad to take a walk about over your farm and the adjoining land, and perhaps I can see you again for a few moments when I return."

"Certainly," Mr. Thornton replied. "There is no objection whatsoever. We are going to Blairville this morning, but we shall be back before noon and I shall be glad to see you then. I fear you have been given some misinformation by the real estate agents. Some of them, by the way, are Northern men who came down here and bought land and when they found they could not make a living on it, they sold it to other land hunters, and I suppose that they made so much in the deal that they stayed right here as real estate agents. They are great advertisers; but I reckon our Southern real estate men can just about keep even. The agent who was out here last spring told me he showed one Northern man a farm for $12 an acre and he was afraid to buy. Then he took him into another county and showed him a poorer farm for $45 and he bought that at once.

"The road there runs out through the fields. Our land runs back to the other public road and beyond that is the farm I told you of where the saw mill is running. I've got some pretty good cowpeas you'll pass by. I haven't got them off the racks yet."

Percy found the cowpea hay piled in large shocks over tripods made of short stout poles which served to keep the hay off the ground to some extent, and this permitted the cowpeas to be cured in larger piles and with less danger of loss from molding.

"I find that the soil on your farm and on the other farm is very generally acid," said Percy a few hours later when Mr. Thornton asked what he thought of the condititons of farming. "Have you used any lime for improving the soil?"

"Yes, I tried it about ten years ago, and it helped some, but not enough to make it pay. I put ten barrels on about three acres. I thought it helped the corn and wheat a little, and it showed right to the line where I put cowpeas on the land, but I don't think it paid, and it's mighty disagreeable stuff to handle."

"Do you remember how much it cost?" Percy asked.

"Yes, Sir. The regular price was a dollar a barrel, but by taking ten barrels I got the ton for eight dollars; but I'd rather have eight dollars' worth of bone meal."

"I think the lime would be a great help to clover," said Percy.

"Yes, that might be. They tell me that they used to grow lots of clover here; but it played out completely, and nobody sows clover now, except occasionally on an old feed lot which is rich enough to grow anything. It takes mighty good land to grow clover; but cowpeas are better for us. They do pretty well for this old land, only the seed costs too much, and they make a sight of work, and they're mighty hard to get cured. You see they aren't ready for hay till the hot weather is mostly past. If we could handle them in June and July, as we do timothy we'd have no trouble; but we don't get cowpeas planted till June, and September is a poor time for haying."

"It seems to me that clover is a much more satisfactory crop," said Percy. "One can sow clover with oats in the spring, or on wheat land in the late winter, and there is no more trouble with it until it is ready for haying about fifteen months later, unless the land is weedy or the clover makes such a growth the first fall that we must clip it to prevent either the weeds or the clover from seeding. This means that when you are planting your ground for cowpeas the next year after wheat or oats, we are just ready to begin harvesting our clover hay; and besides the regular hay crop we usually have some growth the fall before which is left on the land as a fertilizer, and then we get a second crop of clover which we save either for hay or seed. Even after the seed crop is harvested there is usually some later fall growth, and some let the clover stand till it grows some more the next spring and then plow it under for corn."

"I can see that clover would be much better than cowpeas if we could grow it; but, as I said, it's played out here. Our land simply won't grow it any more. Not having to plow for clover would save a great deal of the work we must do for our cowpeas."

"Some of our farmers follow a three-year rotation and plow the ground only once in three years," said Percy. "They plow the ground for corn, disk it the next spring when oats and clover are seeded, and then leave the land in clover the next year. In that way they regularly harvest four crops, including the two clover crops, from only one plowing; and in exceptional seasons I have known an extra crop of clover hay to be harvested in the late fall on the land where the oats were grown.

"In regard to the lime question," Percy continued, "I wonder if you know of the work the Pennsylvania Experiment Station has been doing with the use of ground limestone in comparison with burned lime."

"No, I never heard of ground limestone being used. I supposed it had to be burned. I should think it would be very expensive to grind limestone."

"No, it costs much less to grind it than to burn it," Percy replied. "Mills are used for grinding rock in cement manufacture, and the rock phosphate and bone meal must all be ground before using them either for direct application or for the manufacture of acidulated fertilizers; and limestone is not so hard to grind as some other rocks. Furthermore it does not need to be so very finely ground. If fine enough so that it will pass through a sieve with ten meshes to the inch it does very well. That you see would be a hundred meshes to the square inch; and, of course, a great deal of it will be much finer than that. In fact the ground limestone used in the Pennsylvania experiments was only fine enough so that about ninety per cent. of it would pass a sieve with ten meshes to the inch, and yet the limestone gave decidedly better results than the burned lime, and it is not nearly so disagreeable to handle. Besides this, the ground limestone is much less expensive. It can be obtained at most points in Illinois for about a dollar and fifty cents a ton."

"A dollar and fifty cents a ton!" exclaimed Mr. Thornton. "Well, that is cheap, but how about the freight and the barrels and bags? Freight is a big item with us."

"The dollar and fifty cents includes the freight," was the reply.

"Includes the cost and the freight both?"

"Yes, and the Illinois farmers have it shipped in bulk, so there is no expense for barrels or bags. Of course the supplies of both coal and limestone are very abundant, and with a well-equipped plant the actual cost of grinding does not exceed twenty-five cents a ton. The original cost of the material ground and on board cars at the works varies from about sixty cents to one dollar a ton, and this leaves a very fair margin of profit.

"The men who furnish the ground limestone realize that very large quantities of it are needed if the soils of Illinois are to be kept fertile, and they also realize that the ultimate prosperity of the country depends upon agricultural prosperity. Their far-sightedness and patriotism combine to lead them to try to sell carloads of limestone instead of tons of burned lime. As a matter of fact five or ten dollars profit on a car of limestone, the use of which in large quantities is thus made possible in systems of positive soil improvement, is very much better for all concerned than a profit of half that much on a single ton of burned lime which is used as a soil stimulant in systems of soil exhaustion."

"It is certainly true," said Mr. Thornton, "that all other great industries depend upon agriculture, directly or indirectly. I have thought of it many times. It seems to me that fishing is about the only exception of importance."

Mr. Thornton requested that Percy remain for lunch in order that they might return to the field to let him see the soil acidity tests made.

CHAPTER XIII

WHY PERCY WENT TO COLLEGE

"I AM interested to know where you learned these things about acid soils and lime and limestone," said Mr. Thornton.

"Mostly in the agricultural college," replied Percy, "but much of the information really comes from the investigations that are conducted by the experiment stations. For example, the best information the world affords concerning the comparative value of burned lime and ground limestone is furnished by the Pennsylvania Agricultural Experiment Station. Those experiments have been carried on continuously since 1882, and the results of twenty years' careful investigations have recently been published. A four-year rotation of crops was practiced, including corn, oats, wheat, and hay, the hay being clover and timothy mixed. With every crop the limestone has given better results than the burned lime. In fact the burned lime seems to have produced injurious results of late years, and the analysis of the soil shows that there has been large loss of humus and nitrogen where the burned lime has been used, the actual loss being equivalent to the destruction of more than two tons of farm manure per acre per annum."

"Well, we surely need this information," said Mr. Thornton. "I have always supposed that the teachers in the agricultural college knew little or nothing of practical farming."

"I did not go to college to learn practical farming, if we mean by that the common practice of agriculture," replied Percy. "I already knew what we call practical farming; that is, how to do the ordinary farm work, including such operations as plowing, planting, cultivating, and harvesting; but it seems to me, Mr. Thornton, that this sort of practical farming has resulted in practical ruin for most of these Eastern lands. The fact is there is a side to agriculture that I knew almost nothing about as a so-called practical farmer, and I am coming to believe that what we commonly call practical farming is often the most impractical farming,—certainly this is true if it ultimately results in depleted and abandoned lands. The truly practical farmer is the man who knows not only how to do, but also what to do and why he does it. The Simplon railroad tunnel connecting Switzerland with Italy is twelve miles long,—the longest in the world. It was dug from the two ends, but under the mountain, six miles from either end, the two holes came together exactly, within a limit of error of less than six inches, and made one continuous tunnel twelve miles long. Now, this was not all accomplished by the practical men who knew how to handle a spade in digging a ditch. The work was controlled by science, and it was known in advance what the results would be. I do not mean that it was known how hard the digging would be, nor how much trouble would be caused by caving or by water; but it was known that if the practical work was done, the final outcome would be successful.

"I think it is even more important that we understand enough of the sciences which underlie the practice of agriculture so we may know in advance that when the practical farm work is done the soil will be richer and better rather than poorer and less productive because of our impractical farming.

"As I said, I did not go to the agricultural college to learn the practice or art of farming; I went to learn the science of agriculture; but, as a matter of fact, I found the college professor knew about as much of practical agriculture as I did and a great deal of science that I did not know. I found that the Dean of the college, who is also Director of the Experiment Station, had been born and raised on the farm, had done all kinds of farm work, the same as other farm boys, had gone through an agricultural college, and after his graduation had returned to the farm and remained there for ten years doing his own work with his own hands. He has had as much actual farm experience as you have had, Mr. Thornton, and ten years more than I have had. He was finally called from the farm to become an assistant in the college from which he was graduated, and in a few years he was advanced to head professor in agriculture. About ten years ago he was made dean and director of the agricultural college and experiment station in my own state; and I have been told that he will not recommend any one for a responsible position in an agricultural college unless he has had both farm experience and scientific training. He and most of his associates are owners of farms and would return to them again if they did not feel that they are of more service to agriculture as teachers and investigators."

"I am very glad to know about this," said Mr. Thornton. "Certainly your opinion, based upon such knowledge as you have of your own college, is worth more than all the common talk I have ever heard from those who never saw an agricultural college. I wish you would tell me something more in regard to what crops are made of and about the methods of making land better even while we are taking crops from it every year."

CHAPTER XIV

A LESSON IN FARM SCIENCE

"THE subject is somewhat complicated," Percy replied, "yet it involves no more difficult problems than have been solved in many other lines. The chief trouble is that we have done too little thinking about our own real problems. Even in the country schools we have learned something of banking and various other lines of business, something of the history and politics of this and other countries, something of the great achievements in war, in discovery and exploration, in art, literature, and invention; but we have not learned what our soils contain nor what our crops require. Not one farmer in a hundred knows what chemical elements are absolutely required for the production of our agricultural plants, and one may work hard on the farm from four o'clock in the morning till nine o'clock at night for forty years and still not learn what corn is made of.

"All agricultural plants are composed of ten chemical elements, and the growth of any crop is absolutely dependent upon the supply of these plant food elements. If the supply of any one of these plant food elements is limited, the crop yield will also be limited. The grain and grass crops, such as corn, oats, wheat, and timothy, also the root crops and potatoes, secure two elements from the air, one from water, and seven from the soil.

"The supply of some elements is constantly renewed by natural processes, and iron, one of the ten, is contained in all normal soils in absolutely inexhaustible amount; while other elements become deficient and the supply must be renewed by man, or crop yields decrease and farming becomes unprofitable.

"Matter is absolutely indestructible. It may change its form, but not a pound of material substance can be destroyed. Matter moves in cycles, and the key to the problem of successful permanent agriculture is the circulation of plant food. While some elements have a natural cycle which is amply sufficient to meet all requirements for these elements as plant food, other elements have no such cycle, and it is the chief business of the farmer to make these elements circulate.

"Take carbon, for example. This element is well represented by hard coal. Soft coal and charcoal are chiefly carbon. The diamond is pure crystallized carbon, and charcoal made from pure sugar is pure, uncrystallized carbon. This can easily be made by heating a lump of sugar on a red hot stove until only a black coal remains. Now these different solid materials represent carbon in the elemental form or free state. But carbon may unite with other elements to form chemical compounds, and these may be solids, liquids, gases.

"Thus carbon and sulfur are both solid elements, one black and the other yellow, as generally found. If these two elements are mixed together under ordinary conditions no change occurs. The result is simply a mixture of carbon and sulfur. But, if this mixture is heated in a retort which excludes the air, the carbon and sulfur unite into a chemical compound called carbon disulfid. This compound is neither black, yellow, nor solid; but it is a colorless, limpid liquid; and yet it contains absolutely nothing except carbon and sulfur."

"That seems strange," remarked Mr. Thornton. "Yes, but similar changes are going on about us all the time," replied Percy. "We put ten pounds of solid black coal in the stove and an hour later we find nothing there, except a few ounces of ashes which represent the impurities in the coal."

"Well, the coal is burned up and destroyed, is it not?"

"The carbon is burned and changed, but not destroyed. In this case, the heat has caused the carbon to unite with the element oxygen which exists in the air in the form of a gas, and a chemical compound is formed which we call carbon dioxid. This compound is a colorless gas. This element oxygen enters the vent of the stove and the compound carbon dioxid passes off through the chimney. If there is any smoke, it is due to small particles of unburned carbon or other colored substances.

"As a rule more or less sulfur is contained in coal, wood, and other organic matter, and this also is burned to sulfur dioxid and carried into the air, from which it is brought back to the soil in rain in ample amounts to supply all of the sulfur required by plants.

"Everywhere over the earth the atmosphere contains some carbon dioxid and this compound furnishes all agricultural plants their necessary supply of both carbon and oxygen. In other words, these are the two elements that plants secure from the air. The gas, carbon dioxid, passes into the plant through the breathing pores on the under side of the leaves. These are microscopic openings but very numerous. A square inch of a corn leaf may have a hundred thousand breathing pores."

"Now, as we go on, I am especially anxious to get at this question of supply and demand," said Mr. Thornton. "I think I understand about iron and sulfur, and also that these two elements, carbon and oxygen, are both contained in the air in the compound called carbon dioxid, and that this must supply our crops with those two elements of plant food. I'd like to know about the supply. How much is there in the air and how much do the crops require?"

"As you know," said Percy, "the atmospheric pressure is about fifteen pounds to the square inch."

"Yes, I've heard that, I know."

"Well, that means, of course, that there are fifteen pounds of air resting on every square inch of the earth's surface; in other words, that a column of air one inch square and as high as the air goes, perhaps fifty miles or more, weighs fifteen pounds."

"Yes, that is very clear."

"There is only one pound of carbon in ten thousand pounds of ordinary country air. Now, there are one hundred and sixty square rods in an acre, and since there are twelve inches in a foot and sixteen and one-half feet in a rod, it is easy to compute that there are nearly a hundred million pounds of air on an acre, and that the carbon in this amounts to only five tons. A three-ton crop of corn or hay contains one and one-fourth tons of the element carbon; so that the total amount of the carbon in the air over an acre of land is sufficient for only four such crops; while a single crop of corn yielding a hundred bushels to the acre, such as we often raise in Illinois on old feed-lots or other pieces of well treated land would require half of the total supply of carbon contained in the air over an acre. However, the largest crop of corn ever grown, of which there is an established authentic record, was not raised in Illinois, but in the state of South Carolina, in the county of Marlborough, in the year 1898, by Z. J. Drake; and, according to the authentic report of the official committee that measured the land and saw the crop harvested and weighed, and awarded Drake a prize of five hundred dollars given by the Orange Judd Publishing Company,—according to this very creditable evidence, that acre of land yielded 239 bushels of thoroughly aid-dried corn; and such a crop, Mr. Thornton, would require as much carbon as the total amount contained in the air over an acre of land."

"Well, that is astonishing! Then there must be some other source of supply besides the air."

"There is no other direct source from which plants secure carbon; but of course the air is in constant motion. Only one-fourth of the earth's surface is land, and perhaps only one-fourth of this land is cropped, and the average crop is about one-fourth of three tons; so that the total present supply of carbon in the air would be sufficient for about two hundred and fifty years. But as a matter of fact the supply is permanently maintained by the carbon cycle. Thus the carbon of coal that is burned in the stove returns to the air in carbon dioxid; and all combustion of coal and wood, grass and weeds, and all other vegetable matter returns carbon to the atmosphere. All decay of organic matter, as in the fermentation of manure in the pile and the rotting of vegetable matter in the soil, is a form of slow combustion and carbon dioxid is the chief produce of such decay. Sometimes an appreciable amount of heat is developed, as in the steaming pile of stable refuse lying in the barnyard, while the heat evolved in the soil is too quickly disseminated to be apparent.

"In addition to all this, every animal exhales carbon dioxid. The body heat and the animal force or energy are supplied by the combustion of organic food within the body, and here, too, carbon dioxid is the chief product of combustion.

"Thus, as a general average, the amount of carbon removed from the atmosphere by growing plants is no greater than the amount returned to the air by these various forms of combustion or decay. In like manner the supply of combined oxygen is maintained, both carbon and oxygen being furnished to the plant m the carbon dioxid.

"As a matter of fact, the air consists very largely of oxygen and nitrogen, both in the free state, but in this form these elements cannot be utilized in the growth of agricultural plants. The only apparent exception to this is in case of legume crops, such as clover, alfalfa, peas, beans, and vetch, which have power to utilize the free nitrogen by means of their symbiotic relationship with certain nitrogen-fixing bacteria which live, or may live, in tubercles on their roots.

"Carbon and oxygen constitute about ninety per cent. of the dry matter of ordinary farm crops, and with the addition of hydrogen very important plant constituents are produced; such as starch, sugar, fiber, or cellulose, which constitute the carbohydrate group. As the name indicates, this group contains carbon, hydrogen, and oxygen, the last two being present in the same proportion as in water.

"Water is composed of the two elements, hydrogen and oxygen, both of which are gases in the free state. Water is taken into the plant through the roots and decomposed in the leaves in contact with the carbon dioxid under the influence of sunlight and the life principle. The oxygen from the water and part of that from the carbon dioxid is given off into the air through the breathing pores, while the carbon, hydrogen, and part of the oxygen, unite to form the carbohydrates. These three elements constitute about ninety-five per cent. of our farm crops, and yet every one of the other seven plant food elements is just as essential to the growth and full development of the plant as are these three."

"Then so long as we have air above and moisture below, our crops will not lack for carbon, oxygen, and hydrogen. Is that the summing up of the matter?"

"Yes, Sir," Percy replied.

"And those three elements make up ninety-five per cent. of our farm crops. Is that correct?"

"Yes, Sir, as an average."

"Well, now it seems to me, if nature thus provides ninety-five per cent. of all we need, we ought to find some way of furnishing the other five per cent. It makes me think of the young wife who told her husband she could live on bread and water, with his love, and he told her that if she would furnish the bread he'd skirmish around and get the water. But, say, did that South Carolina man use any fertilizer for that immense crop of corn?"

"Some fertilizer, yes. He applied manure and fertilizer from February till June. In all he applied 1000 bushels (about 30 tons) of farm manure, 600 bushels of whole cotton seed, 900 pounds of cotton seed meal, 900 pounds of kainit, 1100 pounds of guano, 200 pounds of bone meal, 200 pounds of acid phosphate, and 400 pounds of sodium nitrate."

"I would also like to know the facts about this nitrogen business," said Mr. Thornton. "I've understood that one could get some of it from the air, and I would much rather get it that way than to buy it from the fertilizer agent at twenty cents a pound. Cowpeas don't seem to help much, and we don't have the cotton seed, and we never have sufficient manure to cover much land."

"It is a remarkable fact," said Percy, "that of the ten essential elements of plant food, nitrogen is the most abundant, measured by crop requirements, and at the same time the most expensive. The air above an acre of land contains enough carbon for a hundred bushels of corn per acre for two years, and enough nitrogen for five hundred thousand years; and yet the nitrogen in commercial fertilizers costs from fifteen to twenty cents a pound. At commercial prices for nitrogen, every man who owns an acre of land is a millionaire.

"You mean he has millions in the air," amended Mr. Thornton.

"Yes, that is the better way to put it," Percy admitted, "but the fact is he can not only get this nitrogen for nothing by means of legume crops, but he is paid for getting it, because those crops are profitable to raise for their own value. Clover, alfalfa, cowpeas, and soy beans are all profitable crops, and they all have power to use the free nitrogen of the air.

"There are a few important facts to be kept in mind regarding nitrogen:

"A fifty-bushel crop of corn takes 75 pounds of nitrogen from the soil. Of this amount about 50 pounds are in the grain, 24 pounds are in the stalks, and 1 pound in the cobs. A fifty-bushel crop of oats takes 48 pounds of nitrogen from the soil, 33 pounds in the grain, and 15 in the straw. A twenty-five bushel crop of wheat also takes 48 pounds of nitrogen from the soil, 36 pounds in the grain and 12 in the straw.

"These amounts will vary to some extent with the quality of the crops, just as the weight of a bushel of wheat varies from perhaps 56 to 64 pounds, although as an average wheat weighs 60 pounds to the bushel."

"You surely remember figures well," remarked Mr. Thornton as he made some notations.

"It is easy to remember what we think about much and often," said Percy; "as easy to remember that a ton of cowpea hay contains 43 pounds of nitrogen as that Blairville is 53 miles from Richmond."

"I have added those figures together," continued Mr. Thornton, "and I find that the three crops, corn, oats, and wheat, would require 171 pounds of nitrogen. Now suppose we raise a crop of cowpeas the fourth year, how much nitrogen would be added to the soil in the roots and stubble?"

"Not any."

"Do you mean to say that the roots and stubble of the cowpeas would add no nitrogen to the soil? Surely that does not agree with the common talk."

"It is even worse than that," said Percy. "The cowpea roots and stubble would contain less nitrogen than the cowpea crop takes from a soil capable of yielding thirty bushels of corn or oats. Only about one-tenth of the nitrogen contained in the cowpea plant is left in the roots and stubble when the crop is harvested. Suppose the yield is two tons per acre of cowpea hay! Such a crop would contain about 86 pounds of nitrogen, and about 10 pounds of nitrogen per acre would be left in the roots and stubble."

"Well, that wouldn't go far toward replacing the 171 pounds removed from the soil by the corn, oats, and wheat, that's sure," was Mr. Thornton's comment.

"It is worse than that," Percy repeated. "Land that will furnish 48 pounds of nitrogen for a crop of oats or wheat will furnish more than 10 pounds for a crop of cowpeas. At the end of such a four-year rotation such a soil would be about 200 pounds poorer in nitrogen per acre than at the beginning, if all crops were removed and nothing returned."

"How much would it cost to put that nitrogen back in commercial fertilizer?" asked Mr. Thornton.

"That depends, of course, upon what kind of fertilizer is used."

"Well, most people around here who use fertilizer buy what the agent calls two-eight-two, and its costs about one dollar and fifty cents a hundred pounds; but it can be bought by the ton for about twenty-five dollars."

"'Two-eight-two' means that the fertilizer is guaranteed to contain two per cent. of ammonia, eight per cent. of available 'phosphoric acid,' and two per cent. of potash."

"Ammonia is the same as nitrogen, is it not?"

"No, it is not the same," replied Percy. "Ammonia is a compound of nitrogen and hydrogen. In order to have a clear understanding of the relation between ammonia and nitrogen we only need to know the combining weights of the elements. The smallest particle of an element is called an atom. Hydrogen is the lightest of all the elements and the weight of the hydrogen atom is used as the standard or unit for the measure of all other atomic weights; thus the atom of hydrogen weighs one."

"One what?" interrupted Mr. Thornton.

"No one knows," replied Percy. "The atom is extremely small, much too small to be seen with the most powerful microscope; but you know all things are relative and we always measure one thing in terms of another. We say a foot is twelve inches and an inch is one-twelfth of a foot, and there we stop with a definition of each expressed in terms of the other, and both depending upon an arbitrary standard that somebody once adopted; and yet, while the foot is known in most countries, it is rare that two countries have exactly the same standard for this measure of length.

"We do not know the exact weight of the hydrogen atom, but we do know its relative weight. If the hydrogen atom weighs one then other atomic weights are as follows:

12 for carbon 14 for nitrogen 16 for oxygen 24 for magnesium 31 for phosphorus 32 for sulfur 39 for potassium 40 for calcium 56 for iron

"This means that the iron atom is fifty-six times as heavy as the hydrogen atom. These atomic weights are absolutely necessary to a clear understanding of the compounds formed by the union or combination of two or more elements.

"One other thing is also necessary. That is to keep in mind the number of bonds, or hands, possessed by each atom. The atom of hydrogen has only one hand, and the same is true of potassium. Each atom of oxygen has two hands; so that one oxygen atom can hold two hydrogen atoms in the chemical compound called water (H-O-H or H20). Other elements having two-handed atoms are magnesium and calcium. Strange to say, the sulfur atom has six hands but sometimes uses only two, the others seemingly being clasped together in pairs. I will write it out for you, thus:

Hydrogen sulfid: H-S-H or H2S

Sulfur dioxid: O=S=0 or S02

"The carbon atom has four hands, and atoms of nitrogen and phosphorus have five hands, but sometimes use only three. Thus, in the compound called ammonia, one atom of nitrogen always holds three atoms of hydrogen; so, if you buy seventeen pounds of ammonia you would get only fourteen pounds of nitrogen and three pounds of hydrogen. This means that, if the two-eight-two fertilizer contains two per cent. of ammonia, it contains only one and two-thirds per cent. of the actual element nitrogen, and a ton of such fertilizer would contain thirty-three pounds of nitrogen. In other words it would take six tons of such fertilizer to replace the nitrogen removed from one acre of land in four years if the crop yields were fifty bushels of corn and oats, twenty-five bushels of wheat, and two tons of cowpea hay."

"Six tons! Why, that would cost a hundred and fifty dollars! Well, well, I thought I knew we couldn't afford to keep up our land with commercial fertilizer; but I didn't think it was that bad. Almost forty dollars an acre a year!"

"It need not be quite that bad," said Percy. "You see this two-eight-two fertilizer contains eight per cent. of so-called 'phosphoric acid' and two per cent. of potash, and those constituents may be worth much more than the nitrogen; but, so far as nitrogen is concerned, the two hundred pounds would cost from thirty to forty dollars in the best nitrogen fertilizers in the market, such as dried blood or sodium nitrate."

"Well, even that would be eight or ten dollars a year per acre, and that is as much as the land is worth, and this wouldn't include any other plant food elements, such as 'phosphoric acid' and potash."

"No, that much would be required for the nitrogen alone if bought in commercial form. I understand that the farmers who use this common commercial fertilizer, apply about three hundred pounds of it to the acre perhaps twice in four years. That would cost about eight dollars for the four years, and the total nitrogen applied in the two applications would amount to 10 pounds per acre."

"It is not quite correct to call 'phosphoric acid' and potash plant food elements. They are not elements but compounds."

"Like ammonia, which is part nitrogen and part hydrogen?"

"The problem is somewhat similar, but not just the same," Percy replied. "These compounds contain oxygen and not hydrogen."

"Well, I understand that both oxygen and hydrogen are furnished by natural processes, the oxygen from carbon dioxid in the carbon cycle, and the hydrogen from the water which falls in rain."

"That is all true, but you really do not buy the hydrogen or oxygen. While they are included in the two-eight-two guarantee, the price is adjusted for that. Thus the cost of nitrogen would be just the same whether you purchase the fertilizer on the basis of seventeen cents a pound for the actual element nitrogen, or fourteen cents a pound for the ammonia."

"Yes, I see how that might be, but I don't see why the guarantee should be two per cent. of ammonia instead of one and two-thirds per cent. of nitrogen, when the nitrogen is all that gives it value."

"There is no good reason for it," said Percy. "It is one of those customs that are conceived in ignorance and continued in selfishness. It is very much simpler to consider the whole subject on the basis of actual plant food elements, and I am glad to say that many of the state laws already require the nitrogen to be guaranteed in terms of the actual element, a few states now require the phosphorus and potassium also to be reported on the element basis."

"That is hopeful, at least," said Mr. Thornton. "Now, if I am not asking too many questions or keeping you here too long, I shall be glad to have you explain two more points that come to my mind: First, how much of that two hundred pounds of nitrogen can I put back in the manure produced on the farm; and, second, just what is meant by potash and phosphoric acid?"

Percy made a few computations and then replied: "If you sell the wheat; feed all the corn, oats, and cowpea hay and half of the straw and corn fodder, and use the other half for bedding; and, if you save absolutely all of the manure produced, including both the solid and liquid excrement; then it would be possible to recover and return to the land about 173 pounds of nitrogen during the four years, compared with the 200 pounds taken from the soil."

"I can't understand that," said Mr. Thornton. "How can that be when one of the crops is cowpeas?"

"In average live-stock and dairy farming," Percy continued, "about one-fourth of the nitrogen contained in the food consumed is retained in the milk and animal growth, and you can make the computations for yourself. It should be kept in mind, moreover, that much of the manure produced on the average farm is wasted. More than half of the nitrogen is in the liquid excrement, and it is extremely difficult to prevent loss of the liquid manure. There is also large loss of nitrogen from the fermentation of manure in piles; and when you smell ammonia in the stable, see the manure pile steaming, or colored liquid soaking into the ground beneath, or flowing away in rainy weather, you may know that nitrogen is being lost. How many tons of manure can you apply to your land under such a system of farming as we have been discussing?"

"Well, I've figured a good deal on manure," was the reply, "and I think with four fields producing such crops as you counted on, that I could possibly put ten or twelve tons to the acre on one field every year."

"That would return from 100 to 120 pounds of nitrogen;" said Percy, "instead of the 173 pounds possible to be returned if there is no loss. There are three methods that may be used to reduce the loss of manure: One of these is to do the feeding on the fields. Another is to haul the manure from the stable every day or two and spread it on the land. The third is to allow the manure to accumulate in deep stalls for several weeks, using plenty of bedding to absorb the liquid and keep the animals clean, and then haul and spread it when convenient."

"I'm afraid that last method would not do at all for the dairy farmer," said Mr. Thornton. "You see we have to keep things very clean and in sanitary condition."

"Most often the cleanest and most sanitary method the average farmer has of handling the manure in dairying," said Percy, "is to keep it buried as much as possible under plenty of clean bedding; and one of the worst methods is to overhaul it every day by 'cleaning' the stable, unless you could have concrete floors throughout, and flush them well once or twice a day, thus losing a considerable part of the valuable excrement. If you allow the manure to accumulate for several weeks at a time, it is best to have sufficient room in the stable or shed so that the cows need not be tied. If allowed to run loose they will find clean places to lie down even during the night.

"In case of horses, the manure can be kept buried for several weeks if some means are used to prevent the escape of ammonia.

Cattle produce what is called a 'cold' manure, while it is called 'hot' from horses because it decomposes so readily. One of the best substances to use for the prevention of loss of ammonia in horse stables is acid phosphate, which has power to unite with ammonia and hold it in a fixed compound. About one pound of acid phosphate per day for each horse should be sprinkled over the manure. Of course the phosphorus contained in the acid phosphate has considerable value for its own sake, and care should be taken that you do not lose more phosphorus from the acid phosphate applied than the value of all the ammonia saved by this means. Porous earth floors may absorb very considerable amounts of liquid from wet manure lying underneath the dry bedding, and the acid phosphate sometimes injures the horses' feet; so that, as a rule, it is better to clean the horse stables every day and supply phosphorus in raw phosphate at one-fourth of its cost in acid phosphate."

"Before we leave the nitrogen question," said Mr. Thornton, "I want to ask if you can suggest how we can get enough of the several million dollars' worth we have in the air to supply the needs of our crops and build up our land?"

"Grow more legumes, and plow more under, either directly or in manure."

"That sounds easy, but can you suggest some practical system?"

"I think so. I know too little of your conditions to think I could suggest the best system for you to adopt; but I can surely suggest one that will supply nitrogen for such crop yields as we have considered: Suppose we change the order of the crops and grow wheat, corn, oats, and cowpeas, and grow clover with the wheat and oats, plowing the clover under in the spring as green manure for corn and cowpeas. If necessary to prevent the clover or weeds from producing seed, the field may be clipped with the mower in the late summer when the clover has made some growth after the wheat and oats have been removed. Leave this season's growth lying on the land. As an average it should amount to more than half a ton of hay per acre. The next spring the clover is allowed to grow for several weeks. It should be plowed under for corn on one field early in May and two or three weeks later the other field is plowed for cowpeas. The spring growth should average nearly a ton of clover hay per acre. In this way clover equivalent to about three tons of hay could be plowed under. Clover hay contains 40 pounds of nitrogen per ton; so this would supply about 120 pounds of nitrogen in addition to the 173 pounds possible to be supplied in the manure. This would make possible a total return of 293 pounds, while we figured some 200 pounds removed. Of course if you save only 100 pounds in the manure the amount returned would be reduced to 220 pounds."

"There are two questionable points in this plan," said Mr. Thornton, " one is the impossibility, or at least the difficulty, of growing clover on this land. The other point is, How much of that 120 pounds of nitrogen returned in the clover is taken from the soil itself? I remember you figured 86 pounds of nitrogen in two tons of cowpea hay, but you also assumed that about 29 pounds of it would be taken from the soil."

"Yes, that is true," Percy replied, "at least 29 pounds and probably more. You see the cowpeas grow during the same months as corn and on land prepared in about the same manner. If the soil will furnish 75 pounds of nitrogen to the corn crop, and 48 pounds to the oats and wheat, it would surely furnish 29 pounds to the cowpeas. Of course this particular amount has no special significance, but the other definite amounts removed in corn, oats, and wheat aggregate 171 and the 29 pounds were added to make the round 200 pounds. Perhaps 210 pounds would be nearer the truth, in which case the soil would furnish about half as much nitrogen to the cowpea crop as to the corn crop. This is reasonable considering that corn is the first crop grown after the manure is applied. You will remember that only one-tenth of the total nitrogen of the cowpea plant remains in the roots and stubble?"

"Yes, that's what we figured on."

"The cowpea is an annual plant. It is planted, produces its seed, and dies the same season. It has no need to store up material in the roots for future use. Consequently the substance of the root is largely taken into the tops as the plan approaches maturity. It is different with the clover plant. This is a biennial with some tendency toward the perennial plant. It lives long and develops an extensive root system, and its stores up material in the roots during part of its life for use at a later period. About one-third of the total nitrogen content of the clover plant is contained in the roots and stubble. This means that the roots and stubble of a two-ton crop of clover would contain about forty pounds of nitrogen, or more than we assumed was taken from the soil by the cowpeas. But there is still another point in favor of the clover. The cowpeas make their growth during the summer months when nitrification is most active, whereas the clover growth we have counted on occurs chiefly during the fall and spring when nitrification is much less active, consequently the clover probably takes even a larger proportion of its nitrogen from the air than we have counted on."

"That is rather confusing," said Mr. Thornton, "you say the cowpea grows when nitrification is most active, and yet you say that it takes less nitrogen from the air than clover. Isn't that somewhat contradictory?"

"I think not," said Percy." Let me see.—Just what do you understand by nitrification?"

"Getting nitrogen from the air, is it not?"

"No, no. That explains it. Getting nitrogen from the air is called nitrogen fixation. This action is carried on by the nitrogen-fixing bacteria, such as the clover bacteria, the soy bean bacteria, the alfalfa bacteria, which, by the way, are evidently the same as the bacteria of sweet clover, or mellilotus. Then we also have the cowpea bacteria, and these seem to be the same as the bacteria of the wild partridge pea, a kind of sensitive plant with yellow flowers, and a tiny goblet standing upright at the base of each compound leaf,—the plant called Cassia Chamaecrista by the botanist."

"Nitrification is an altogether— "

"Well, I declare! Excuse me, Sir, but that's Charlie calling the cows. Scotts, I don't see where the time has gone! You'll excuse me, Sir, but I must look after separating the cream. You will greatly oblige me, Mr. Johnston, if you will have dinner with us and share our home to-night. In addition to the pleasure of your company, I confess that I am mightily interested in this subject; and I would like especially to get a clear understanding of that nitrification process, and we've not had time to discuss the potash and 'phosphoric acid,' which I know cost some of our farmers a good part of all they get for their crops, and still their lands are as poor as ever."

"I appreciate very much your kind invitation, Mr. Thornton. I came to you for correct information regarding the agricultural conditions here, and you were very kind and indulgent to answer my blunt questions, even concerning your own farm practice and experience. I feel, Sir, that I am already greatly indebted to you, but it will certainly be a great pleasure to me to remain with you tonight."

For more than two hours they had been standing, leaning, or sitting in a field beside a shock of cowpea hay, Percy toying with his soil auger, and Mr. Thornton making records now and then in his pocket note book.

CHAPTER XV

COEDUCATION

PERCY took a lesson in turning the cream separator and after dinner Mrs. Thornton assured him that she and her sister were greatly disappointed that they had not been permitted to hear the discussion concerning the use of science on the farm.

"We have never forsaken our belief that these old farms can again be made to yield bountiful crops," she said, "as ours did for so many years under the management of our ancestors. 'Hope springs eternal in the human breast.' I stop with that for I do not like the rest of the couplet. We can see that some marked progress has been made under my husband's management, although he feels that it is very slow work building up a run-down farm. But he has raised some fine crops on the fields under cultivation,—as much as ten barrels of corn to the acre, have you not, Dear?" she asked.

"Yes, fully that much, but even ten barrels per acre on one small field is nothing compared to the great fields of corn Mr. Johnston raises in the West. and it makes a mighty small show here on a nine-hundred-acre farm, most of which hasn't been cropped for more than twenty years; and even then it was given up because the negro tenants couldn't raise corn enough to live on.

"I've talked some with the fertilizer agents, but they don't know much about fertilizers, except what they read in the testimonials published in the advertising booklets. I have had some good help from the agricultural papers, but most that is written for the papers doesn't apply to our farm, and it's so indefinite and incomplete, that

I've just spent this whole evening asking Mr. Johnston questions; and I haven't given him a chance to answer them all yet."

"I am sure you have not asked more questions this afternoon than I did this forenoon," Percy remarked; "and all your answers were based on authentic history or actual experience, while my answers were only what I have learned from others."

"Well, if we were more ready to learn from others, it would be better for all of us," said Mr. Thornton. "Experience is a mighty dear teacher and, even if we finally learn the lesson, it may be too everlasting late for us to apply it. Now we all want to learn about that process called nitrification."

"It is an extremely interesting and important process," said Percy. "It includes the stages or steps by which the insoluble organic nitrogen of the soil is converted into soluble nitrate nitrogen, in which form it become available as food for all of our agricultural plants."

"Excepting the legumes?" asked Mr. Thornton.

"Excepting none," Percy replied. "The legume plants, like clover, take nitrogen from the soil so far as they can secure it in available form, and in this respect clover is not different from corn. The respect in which it is different is the power of clover to secure additional supplies of nitrogen from the air when the soil's available supply becomes inadequate to meet the needs of the growing clover. If the conditions are suitable for nitrogen-fixation, then the growth of the legume plants need not be limited by lack of nitrogen; whereas, nitrogen is probably the element that first limits the growth and yield of all other crops on your common soils."

"Now, what do you think of that, Girls? With millions of dollars' worth of nitrogen in the air over every acre, our crops are poor just because we don't use it. I wish you would tell me something about the suitable conditions for nitrogen-fixation, Mr. Johnston. You understand, Girls, that nitrogen-fixation is simply getting nitrogen from the inexhaustible supply in the air by means of little microscopic organisms called bacteria, which live in little balls called tubercles attached to the roots of certain plants called legumes, like cowpeas and clover. Corn and wheat and such crops can't get this nitrogen. Now, Mr. Johnston is telling about nitrification, a process which is entirely different from nitrogen-fixation. Excuse me, Mr. Johnston, but I wanted to make this plain to Mrs. Thornton and Miss Russell."

"I am glad you did so," Percy replied. "As I was saying, nitrification has no connection whatever with the free nitrogen of the air.

"All plants take their food in solution; that is, the plant food taken from the soil must be dissolved in the soil water or moisture. Of the essential elements of plant food, seven are taken from the soil through the roots into the plant. These seven do not include those of which water itself is composed. Now, these seven plant food elements exist in the soil almost exclusively in an insoluble form. In that condition they are not available to the plant for plant food; and it is the business of the farmer to make this plant food available as fast as is needed by his growing crops.

"The nitrogen of the soil exists in the organic matter; that is, in such materials as plant roots, weeds, and stubble, that may have been plowed under, or any kind of vegetable maker incorporated with the soil, including all sorts of crop residues, green manures, and the common farm fertilizers from the stables. When these organic materials are decomposed and disintegrated to such an extent that their structure is completely destroyed, the resulting mass of partially decayed black organic matter is called humus. The nitrogen of the soil is one of the constituents of this humus or other organic matter. It is not contained in the mineral particles of the soil. On the other hand the other six elements of plant food are contained largely in the mineral part of the soil, as the clay, silt, and sand. thus the iron, calcium, magnesium, and potassium, all of which are called abundant elements, are contained in the mineral matter, and usually in considerable amounts, while they are found in the organic matter in very small proportion. The phosphorus and sulfur are found in very limited quantities in most soils, but they are present in both organic and mineral form.

"Practically the entire stock or store of all of the elements in the soil is insoluble and consequently unavailable for the use of growing plants; and, as I said, some of the chief plans and efforts of the farmer should be directed to the business of making plant food available.

"The nitrogen contained in the insoluble organic matter of the soil is made soluble and available by the process called nitrification. Three different kinds of bacteria are required to bring about the complete change."

"Are these bacteria different from the nitrogen fixing bacteria?" asked Mr. Thornton.

"Entirely different," Percy replied, "and there are three distinct kinds, one for each of the three steps in the process.

"The first may be called ammonia bacteria. They have power to convert organic nitrogen into ammonia nitrogen; that is, into the compound of nitrogen and hydrogen; and this step in the process is called ammonification.

"The other two kinds are the true nitrifying bacteria. One of them converts the ammonia into nitrites, and the other changes the nitrites into nitrates. These two kinds are known as the nitrite bacteria and the nitrate bacteria.

"Technically the last two steps in the process are nitrification proper; but, speaking generally, the term nitrification is used to include the three steps, or both ammonification and nitrification proper.

"Now, the nitrifying bacteria require certain conditions, otherwise they will not perform their functions. Among these essential conditions are the presence of moisture and free oxygen, a supply of carbonates, certain food materials for the bacteria themselves, and a temperature within certain limits.

"You may remember, Mr. Thornton, that more soil nitrogen is made available for cowpeas during the summer weather than for clover during the cooler fall and spring?"

"Yes, I remember that distinction."

"I declare," said Miss Russell, "Tom talks as though he had been there and seen the things going on. I haven't seen you using any microscope."

"Well, I tell you, I've mighty near seen 'em," was the reply. "Mr. Johnston makes everything so plain that I can mighty near see what he saw when he looked through the microscope."

"I greatly enjoyed my microscopic work," said Percy, "and still more the work in the chemical laboratory where we finally learned to analyze soils, to take them apart and see what they contain,—how much nitrogen how much phosphorus, how much limestone, or how much soil acidity, which means that limestone is needed. Then I also enjoyed the work in the pot-culture laboratory, where we learned not to analyze but to synthesize; that is, to put different materials together to make a soil. Thus, we would make one soil and put in all of the essential plant food elements except nitrogen, and another with only phosphorus lacking, and still another with both nitrogen and phosphorus present, and all of the other essential elements provided, except potassium, or magnesium, or iron. These prepared soils were put in glass jars having a hole in the bottom for drainage, and then the same kind of seeds were planted in each jar or pot. Some students planted corn, others oats or wheat or any kind of farm seeds. I grew rape plants in one series of pots, and I have a photograph with me which shows very well that all of the plant food elements are essential.

"You see one pot contained no plant food and one was prepared with all of the ten essential elements provided. Then the other pots contained all but one of the necessary soil elements, as indicated in the photograph."

"Why, I never saw anything like that," said Mrs. Thornton.

"But I have many a time," said her husband, "right here on this old farm; I don't know what's lacking, of course, but some years I've thought most everything was lacking. But, according to this pot-culture test, you can't raise any crops if just one of these ten elements is lacking, no matter how much you have of the other nine; and it seems to make no difference which one is lacking, you don't get any crop. Is that the fact, Mr. Johnston?"

One pot with no plant food, and one with all the essential elements provided, and still others with but one element lacking. All planted the same day and cared for alike.

"Yes, Sir," Percy replied. "Where all of the elements are provided, a fine crop is produced, but in each case where a single element is omitted that is the only difference, and in some cases the result is worse than where no plant food is supplied. It seems to hurt the plant worse to throw its food supply completely out of balance than to leave it with nothing except what it draws from the meager store in the seed planted. Of course all the pots were planted with the same kind of seed at the same time, and they were all watered uniformly every day."

"Those results are very striking, indeed," said Miss Russell," but I suppose one would never see such marked differences under farm conditions?"

"Only under unusual or abnormal conditions," Percy replied, "but the fact is that as a very general rule our crop yields are limited chiefly because the supply of available plant food is limited. Sometimes the clover crop is a complete failure on untreated land, while it lives and produces a good crop if the soil is properly treated; and in such cases the difference developed in the field is just as marked as in the pot-cultures. In general we may set it down as an absolute fact that the productive power of normal land depends primarily upon the ability of the soil to feed the crop.

"I have here a photograph of a corn field on very abnormal soil. They had the negative at the Experiment Station and I secured a print from it, in part because I became interested in a story connected with this experiment field, which our professor of soil fertility reported to us.

"This shows a field of corn growing on peaty swamp land, of which there are several hundred thousand acres in the swamp regions of Illinois, Indiana, and Wisconsin. This peaty soil is extremely rich in humus and nitrogen, well supplied with phosphorus and other elements, except potassium; but in this element it is extremely deficient. This land was drained out at large expense, and produced two or three large crops because the fresh grass roots contained some readily available potassium; but after three or four years the corn crop became a complete failure, as you see from the untreated check plot on the right; while the land on the left, where potassium was applied, produced forty-five bushels per acre the year this photograph was taken, and with heavier treatment from sixty to seventy-five bushels are produced."

"Seventy-five bushels would be fifteen barrels of corn per acre. How's that, Little Wife?" asked Tom.

"It's even more wonderful than the pot culture," replied Mrs. Thornton; "but how much did the potassium cost, Mr. Johnston."

"About three dollars an acre," replied Percy; "but of course the land has almost no value if not treated; and as a matter of fact the three dollars is less than half the interest on the difference in value between this land and our ordinary corn belt land. These peaty swamp lands are to a large extent in scattered areas, and commonly, if a farmer owns some of this kind of land, he also owns some other good land, perhaps adjoining the swamp; but this is not always the case, and was not with the man in the story I mentioned. This man lived a few miles away and his farm was practically all of this peaty swamp land type. He heard of this experiment field and came with his family to see it.

"As he stood looking, first at the corn on the treated and untreated land, and then at his wife and large family of children, he broke down and cried like a child. Later he explained to the superintendent who was showing him the experiments, that he had put the best of his life into that kind of land. 'The land looked rich,' said he,—'as rich as any land I ever saw. I bought it and drained it and built my home on a sandy knoll. The first crops were fairly good, and we hoped for better crops; but instead they grew worse and worse. We raised what we could on a small patch of sandy land, and kept trying to find out what we could grow on this black bogus land. Sometimes I helped the neighbors and got a little money, but my wife and I and my older children have wasted twenty years on this land. Poverty, poverty, always! How was I to know that this single substance which you call potassium was all we needed to make this land productive and valuable? Oh, if I had only known this twenty years ago, before my wife had worked like a slave,—before my children had grown almost to manhood and womanhood, in poverty and ignorance!'"

"Why wasn't the matter investigated sooner?" asked Miss Russell. "Why didn't the government find out what the land needed long before?"

"I am a Yankee," said Percy. "Why have American statesmen ridden back and forth to the national capitol through a wilderness of depleted and abandoned farms in the eastern states for half a century or more before the first appropriation was made for the purpose of agricultural investigation? and why, even now, does not this rich federal government appropriate to the agricultural experiment station in every state a fund at least equal to the aggregate salaries of the congressmen from the same state, this fund to be used exclusively for the purpose of discovering and demonstrating profitable systems of permanent agriculture on every type of soil? Why do we as a nation expend five hundred million dollars annually for the development of the army and navy, and only fifteen millions for agriculture, the one industry whose ultimate prosperity must measure the destiny of the nation?

"Moralists sometimes tell us that the fall of the Babylonian Empire, the fall of the Egyptian Empire, of the Grecian Empire, and the Roman Empire, were all due to the development of pride and immorality among those peoples; whereas, we believe that civilization tends rather toward peace, security, and higher citizenship. Is not the chief explanation for the ultimate and successive fall of those great empires to be found in the exhausted or wasted agricultural resources of the country?

"The land that once flowed with milk and honey might then support a mighty empire, with independent resources sufficient for times of great emergencies, but now that land seems almost barren and supports a few wandering bands of marauding Arabs and villages of beggars.

"The power and world influence of a nation must pass away with the passing of material resources; for poverty is helpless, and ignorance is the inevitable result of continued poverty. Only the prosperous can afford education or trained intelligence.

"Old land is poorer than new land. There are exceptions, but this is the rule. The fact is known and recognized by all America.

"What does it mean? It means that the practice of the past and present art of agriculture leads toward land ruin,—not only in China, where famine and starvation are common, notwithstanding that thousands and thousands of Chinese are employed constantly in saving every particle of fertilizing material, even gathering the human excrements from every house and by-place in village and country, as carefully as our farmers gather honey from their hives; not only in India where starvation's ghost is always present, where, as a rule, there are more hungry people than the total population of the United States; not only in Russia where famine is frequent; but, likewise in the United States of America, the present practice of the art of agriculture tends toward land ruin.

"Nations rise and fall; so does the productive power of vast areas of land. Better drainage, better seed, better implements, and more thorough tillage, all tend toward larger crops, but they also tend toward ultimate land ruin, for the removal of larger crops only hastens soil depletion.

"To bring about the adoption of systems of farming that will restore our depleted Eastern and Southern soils, and that will maintain or increase the productive power of our remaining fertile lands of the Great Central West, where we are now producing half of the total corn crop of the entire world, is not only the most important material problem of the United States; but to bring this about is worthy of, and will require, the best thought of the most influential men of America. Without a prosperous agriculture here there can be no permanent prosperity for our American institutions. While some small countries can support themselves by conducting trade, commerce, and manufacture, for other countries, American agriculture must not only be self-supporting, but, in large degree, agriculture must support our other great industries.

"Without agriculture, the coal and iron would remain in the earth, the forest would be left uncut, the railroads would be abandoned, the cities depopulated, and the wooded lands and water-ways would again be used only for hunting and fishing. Shall we not remember, for example, that the coal mine yields a single harvest—one crop—and is then forever abandoned; while the soil must yield a hundred—yes, a thousand crops, and even then it must be richer and more productive than at the beginning, if those who come after us are to continue to multiply and replenish the earth.

"Even the best possible system of soil improvement, we must admit, is not the absolute and final solution of this, the most stupendous problem of the United States. If war gives way to peace and pestilence to science, then the time will come when the soils of America shall reach the limit of the highest productive power possible to be permanently maintained, even by the general adoption of the most practical scientific methods; and before that limit is reached, if power, progress, and plenty are to continue in our beloved country, there must be developed and enforced the law of the survival of the fittest; otherwise there is no ultimate future for America different from that of China, India, and Russia, the only great agricultural countries comparable to the United States. An enlightened humanity must grant to all the right to live, but the reproduction and perpetuation of the unfit can never be an absolute and inalienable right.

"Under the present laws and customs, a man may spend half his life in the insane asylum or in the penitentiary, and still be the father of a dozen children with degenerate tendencies. There should be no reproduction from convicted criminals, insane persons, and other degenerates. Thieves, grafters, bribers and bribe-takers all belong in the same class, and it should not be left possible for them to reproduce their kind. They are a burden upon the public which the public must bear, but the public is under no obligation to permit their multiplication. The children of such should never become the parents of others. It is a crime against both the child and the public.

"No doubt you will consider this extremely visionary, and so it is; but unless America can see a vision somewhat like this, a population that is doubling three or four times each century, and an area of depleted soils that is also increasing at a rapid rate will combine to bring our Ship of State into a current against which we may battle in vain; for there is not another New World to bring new wealth, new prosperity, and new life and light after another period of 'Dark Ages.'

"Whether we shall ever apply any such intelligence to the possible improvement of our own race as we have in the great improvement of our cattle and corn is, of course, an open question; but to some extent you will agree that the grafter and the insane, like the poet, are born and not made. Of course there are, and always will be, marked variations, mutants, or 'sports,' but, nevertheless, natural inheritance is the master key to the improvement of every form of life; and it is an encouraging fact that some of the states, as Indiana, for example, have already adopted laws looking toward the reduction of the reproduction of convicted degenerates."

CHAPTER XVI

PAST SELF REDEMPTION

"BUT I have rambled far from the subject assigned me," Percy continued.

"That's only because I interrupt and ask so many side questions," replied Mr. Thornton, "but I hope yet to learn more about those 'suitable conditions' for nitrogen-fixation and nitrification. It begins to look as though the nitrogen cycle deviates a good deal from a true circle, and nature seems to need some help from us to make that element circulate as fast as we need it. I confess, too, that this method appeals to me much more than the twenty-cent-a-pound proposition of the fertilizer agent."

"Yes, indeed," added Miss Russell; "and if we had to spend three dollars an acre on this farm our 'Slough of Despond' would be worse than the slough, or swamp, Mr. Johnston has told us about."

"I fear the practical and profitable improvement of an acre of this land is more likely to cost thirty dollars than three," said Percy.

"Oh, for the land's sake!" came the ejaculation.

"Yes, 'for the land's sake,'" repeated Percy; "and for the sake of those who must depend upon the land for their support for all time hereafter."

"How ridiculous! Thirty dollars an acre for the improvement of land that will not bring ten dollars to begin with!"

"It is better to look at the other end of the undertaking," said Percy. "Suppose you invest thirty dollars an acre and in a few years make your ten-dollar land produce as much as our two-hundred dollar land!"

"But, Mr. Johnston; do you realize how much money it would require to expend thirty dollars an acre on nine hundred acres?" continued Miss Russell, with stronger accentuation.

"Twenty-seven thousand dollars," was the simple reply.

"Well, Sir," she said, "you are welcome to this whole farm for ten thousand dollars."

"I am not wishing for it," he answered. "In fact I would not take this farm as a gift, if I were obliged to keep it and pay the taxes and had no other property or source of income."

"That's just the kind of talk I've been putting up to these girls," said Mr. Thornton. "By the time we live and pay about two hundred dollars a year taxes on all this land, I tell you, there is nothing left; and we'd been worse off than we are, except for the sale we made to the railroad company."

"Well, the Russells lived here very well for more than a hundred years," she retorted, "and my grandfather supported one nigger for every ten acres of the farm, but I would like to know any farmers about here who can put thirty dollars an acre, or even ten dollars an acre, back into their soil for improvement."

"The problem is indeed a serious one," said Percy. "Unquestionably much of the land in these older states is far past the point of possible self-redemption under the present ownership. Land from which the fertility has been removed by two hundred years of cropping, until it has ceased to return a living to those who till it, cannot have its fertility restored sufficiently to again make its cultivation profitable, except by making some considerable investment in order to replace those essential elements the supply of which has become so limited as to limit the crop yields to a point where their value is below the cost of production. Even on the remaining productive lands in the North Central States, if we are ever to adopt systems of permanent agriculture, it must be done while the landowners are still prosperous. If the people of the corn belt repeat the history of the Eastern States until their lands cease to return a profit above the total cost of production, then they, too, will have nothing left to invest in the improvement of their lands."

"But their fertility could still be restored by outside capital?" suggested Mr. Thornton. "I know very well that is the only solution of our problem."

"Well, Tom, I would like to know where the outside capital is coming from," said Miss Russell.

"Marry rich," he replied. "Don't make such a blunder as your sister did."

"I fear that Mr. Johnston will suggest that we sell some more land," remarked Mrs. Thornton.

"All right," replied her sister; "and we will sell it to him. If he won't take the whole farm as a gift, we'll cut it to any length he wishes. Do you consider 'Ten Acres Enough,' Mr. Johnston; or would you prefer 'Three Acres and Liberty?' We'll do our best to enable you to enjoy 'The Fat of the Land.' Just tell us how large a farm you want, I know already that you do not want nine hundred acres."

"My dear Miss Russell," said Percy. "This is so sudden"; whereupon Mr. Thornton nearly fell from his chair and Mrs. Thornton laughed heartily at the sister's expense who blushed as she might have done twenty years before.

"However," Percy resumed, "if you should decide to dispose of about half of that seven hundred acres which you use only as a safety bank for most of your two hundred dollars in taxes, please consider me a prospective taker."

"Take her," said Mr. Thornton, and again confusion reigned.

"Tom is so anxious to get rid of his sister-in-law that he reminds me of the man whose mother-in-law died," said Miss Russell. "He was too far from home to return to the usual funeral, and they telegraphed him the sad news and asked if they should embalm, cremate, or bury the remains. He wired back: 'Embalm, cremate, and bury'"

"That matter of outside capital is by no means so substantial as it might seem," said Percy. "It is worth while to consider how little real wealth there would be in America if the remaining rich lands should become impoverished. The railroads would at once cease to pay dividends, and those who are now millionaires in railroad stock would find themselves on the rapid road to poverty. The manufacturer of finished products from the raw materials raised on the farm, the manufacturer of agricultural implements, and the great urban population whose income is from the trade in raw materials and manufactured goods would soon see their wealth shrivel. The great sky scrapers of the cities would be left for the owls and bats to harbor in, if our agricultural lands ceased to yield their great harvests. Meanwhile the farming people would continue to live upon the meager products still produced from the impoverished soil, even though they had no surplus food to ship into the cities. Human labor would replace that of domestic animals on the farm, just as it has done in China and India, in part because man's labor is worth more than that of the beast, when measured only by the amount of food consumed, and in part because a thousand bushels of grain will support five times as many people can be supported for the same time upon the animal products that could be produced by feeding the grain."

"Oh, that is such a gloomy view to take of it," said Miss Russell.

"And all the world loves an optimist," replied Percy laughingly. "Soils do not wear out; there is no poor land; the farms are better and the crops larger than ever before; and we are the people of the world's greatest nation, with an assured future glory which surposses all conception."

"As soon as we get the canal dug," suggested Mr. Thornton.

"Yes, we will surely be able to dig that Panama ditch," said Percy; "and probably our resources will last to cut a gash or two in our own interior, if we don't build too many battle ships. You know Egypt built three great pyramids before her resources became reduced to such an extent that the people required all their energies to secure a living."

CHAPTER XVII

MORE PROBLEMS

"NOW let us give Mr. Johnston a chance to tell us about the nitrogen problem," said Mr. Thornton. "I'm pretty well satisfied with the natural circulation of carbon, oxygen, and hydrogen; but I want to understand all I can of the practical methods of securing and utilizing nitrogen; and we have heard almost nothing about the other six essential elements which the soil must furnish. Let me see.—I think you said that iron, calcium, magnesium and potassium are usually abundant in the soil, while phosphorus and sulfur are very limited."

"Yes, that is the rule under general or average conditions, but it should be stated that the amount of sulfur required by plants is very small as compared with phosphorus, a difference which places a great distinction between them. Besides considerable quantities of sulfur are returned to the air in the combustion of coal and organic matter, and this returns to the soil in rain. The information thus far secured shows that sulfur rarely if ever limits the crop yields under field conditions; and the same may be said of iron, which is required by plants in very small amount and is contained in practically all soils in enormous quantities.

"While normal soils contain abundance of potassium, with about half as much calcium and one-fourth as much magnesium; yet, when measured by crop requirements for plant food, the supplies of these three elements are not markedly different. On the other hand, about 300 pounds of calcium are lost per acre per annum by leaching from good soils in humid climates, compared with about 10 pounds of potasssium and intermediate amounts of magnesium; so that, of these three elements, calcium requires by far the most consideration and potassium the least, even aside from the use of limestone to correct or prevent soil acidity.

"Among the conditions essential for nitrification may be mentioned the presence of free oxygen and limestone; and of course all bacteria require certain food materials, resembling other plants in this respect."

"Are they plants?" asked Mrs. Thornton. "I thought they were tiny little animals."

"No, they are classified as plants," replied Percy; "but the scientists have difficulty with some of the lower organism to decide whether they are plants or animals. The college boys used to say that some animals were plants in the botanical department and animals again when they studied zoology. Orton says it is easy to tell a cow from a cabbage, but impossible to assign any absolute, distinctive character which will divide animal life from plant life.

"The oxygen is essential for nitrification, because that is an oxidation process. That is, it is a kind of combustion, so to speak. The organic matter is oxidized or converted into substances containing more oxygen than in the original form. In ammonification the carbon is separated or divorced from the nitrogen and united with oxygen. Some of the hydrogen of the organic matter remains temporarily with the carbon, and some is held temporarily with the nitrogen in the form of ammonia.

"The nitrite bacteria replace two of the hydrogen atoms in ammonia with one of oxygen, and insert another oxygen atom between the nitrogen and the remaining hydrogen, thus forming nitrous acid; H-O-N=O, or HNO2.

"The nitrate bacteria then cause the direct addition of another oxygen atom, which is held by the two extra bonds of the nitrogen atom, which you will remember is a five-handed atom.

"Thus you will see the absolute need of free oxygen in the nitrification process; and we can control the rate of nitrification to a considerable extent by our methods of tillage. In soils deficient in organic matter, excessive cultivation may still liberate sufficient nitrogen for a fairly satisfactory crop; and the benefits of such excessive cultivation for potatoes and other vegetables is more often due to increased nitrification than to the conservation of moisture, to which it is frequently ascribed by agricultural writers.

"Thus the more we cultivate, the more we hasten the nitrification, oxidation, or destruction of the organic matter or humus of the soil. Where the soil is well supplied with decaying organic matter, we rarely need to cultivate in a humid section like this, except for the purpose of killing weeds.

"The presence of carbonates in the soil is essential for nitrification, because the bacteria will not continue the process in the presence of their own product. Nitrification ceases if the nitrous or nitric acid remains as such; but, in the presence of carbonates such as calcium carbonate (ordinary limestone) or the double carbonate of magnesium and calcium (magnesian limestone, or dolomite), the nitrous acid or nitric acid is converted into a neutral salt of calcium or magnesium, one of these atoms taking the place of two hydrogen atoms and forming, say, calcium nitrate: $Ca(NO3)2$. At the same time the hydrogen atoms take the place of the calcium in limestone ($CaC03$), and form carbonic acid ($H2CO3$), which at once decomposes into water ($H2O$) and carbon dioxid ($CO2$), which thus escapes as a gas into the air or remains in the pores of the soil.

"The fact that nitrification will not proceed in the presence of acid reminds us that only a certain degree of acidity can be developed in sour milk. Here the lactic acid bacteria produce the acid from milk sugar, but the process stops when about seven-tenths of one per cent. of lactic acid has developed. If some basic substance, such as lime, is then added, the acid is neutralized and the fermentation again proceeds.

"In the general process of decay and oxidation of the organic matter of the soil, the nitrogen thus passes through the forms of ammonia, nitrous acid, and nitric acid, and at the same time the carbon passes into various acid compounds, including the complex humic and ulmic acids, and smaller amounts of acetic acid (found in vinegar), lactic acid, oxalic acid (found in oxalic), and tartaric acid (found in grapes). The final oxidation products of the carbon and hydrogen are carbon dioxid and water, which result from the decomposition of the carbonic acid.

"Now the various acids of carbon and nitrogen constitute one of the most important factors in soil fertility. They are the means by which the farmer can dissolve and make available for the growing crops the otherwise insoluble mineral elements, such as iron, calcium, magnesium, and potassium, all of which are contained in most soils in great abundance. These elements exist in the soil chiefly in the form of insoluble silicates. Silicon itself is a four-handed element which bears somewhat the same relation to the mineral matter of the soil as carbon bears to the organic matter. Quartz sand is silicon dioxid (SiO2). Oxygen, which is present in nearly all substances, including air, water, and most solids, constitutes about one-half of all known matter. Silicon is next in abundance, amounting to more than one-fourth of the solid crust of the earth. Aluminum is third in abundance (about seven per cent), aluminum silicate being common clay. Iron, calcium, potassium, sodium, and magnesium, in this order, complete the eight abundant elements, which aggregate about ninety-eight per cent. of the solid crust of the earth.

"It is worth while to know that about two and one-half per cent. of the earth's crust is potassium, while about one-tenth of one per cent. is phosphorus; also that when a hundred bushels of corn are sold from the farm, seventeen pounds of phosphorus, nineteen of potassium, and seven of magnesium are carried away.

"The acids formed from the decaying organic matter not only liberate for the use of crops the mineral elements contained in the soil in abundance, but they also help to make available the phosphorus of raw phosphate, when naturally contained in the soil, as it is to some extent in all soils, or when applied to the soil in the fine-ground natural phosphate from the mines.

"Now the increase or decrease of organic matter in the soil is measured with a very good degree of satisfaction by the element nitrogen, which is a regular constituent of the organic matter of the soil; and you are already familiar, Mr. Thornton, with the amounts of nitrogen contained in average farm manure and in some of our most common crops."

"Yes, Sir, I have some of the figures in my note book and I mean to have them in my head very soon. But, say, that organic matter seems to be a thing of tremendous importance, and I'm sure we've got mighty little of it. I think about the only thing we'll need to do to make this old farm productive again is to grow the vegetation and plow it under. As it decays, it will furnish the nitrogen, and liberate the phosphorus, potassium, calcium, and magnesium; and we may have plenty of all of them just waiting to be liberated."

"That is altogether possible," said Percy; "but it must be remembered that your soil is acid and consequently will not grow clover or alfalfa successfully, or even cowpeas very satisfactorily. A liberal use of ground limestone and large use of clover may be sufficient to greatly improve your soil; but if I am permitted to separate Miss Russell and the Thorntons "—Mr. Thornton's hilarious "Ha, ha" cut Percy short. He crimsoned and the ladies smiled at each other with expressions that revealed nothing whatever.

"Now let me finish," Percy continued, when Mr. Thornton had somewhat subsided. "I say, if I am permitted to separate Miss Russell and the Thorntons from about three hundred acres of their land, I shall certainly wish to know its total content of phosphorus, potassium, magnesium, and calcium, before I make any purchase; and, if you will remember the pot cultures and the peaty swamp land, I think you'd agree with me.

"Well, I shall be mighty glad to know that myself," said Mr. Thornton, "and we shall much appreciate it if you can tell us how to secure that information."

"We can collect some soil to-morrow," Percy replied, "and send it to a chemist for analysis."

"Good," said Mr. Thornton; "now just one more question, and I think I shall sleep better if I have it answered to-night. Just what is meant by potash and phosphoric acid?"

"Potash," said Percy, "is a compound of potassium and oxygen. The proportions are one atom of oxygen and two atoms of potassium, which you may remember are single-handed and weigh thirty-nine, so that seventy-eight of potassium unite with sixteen of oxygen. A better name for the compound is potassium oxid: K20. The

Latin name for potassium is kalium, and K is the symbol used for an atom of that element. If you were to purchase potassium in the form of potassium chlorid, which in the East is often called by the old incorrect name 'muriate of potash,' the salt might be guaranteed to contain a certain percentage of potash, which, however, consists of eighty-three per cent. of potassium and seventeen of oxygen."

"Just what is this potassium chlorid, or 'muriate of potash'?"

"Pure potassium chlorid contains only the two elements, potassium and chlorin."

"But didn't you say that it was guaranteed to contain potash and that potash is part oxygen? Now you say it contains only potassium and chlorin."

"Yes, I am sorry to say, that this is one of those blunders of our semi-scientific ancestors for which we still suffer. The chemist understands that the meaning of the guarantee of potash is the amount of potash that the potassium present in the potassium chlorid could be converted into. The best you can do is to reduce the potash guarantee to potassium by taking eighty-three per cent. of it; or, to be more exact, divide by ninety-four and multiply by seventy-eight, in order to eliminate the sixteen parts of oxygen.

"It may be well to keep in mind that when the druggist says potash he means potassium hydroxid, KOH, a compound of potassium, hydrogen, and oxygen, as the name indicates."

"You mentioned the word chlorin," said Mr. Thornton. "That is another element?"

"Yes, that is a very common element. Ordinary table salt is sodium chlorid: NaCl. Sodium is called natrium in Latin, and Na is the symbol used in English to be in harmony with all other languages, for practically all use the same chemical symbols. Sodium and potassium are very similar elements in some respects, and in the free state they are very peculiar, apparently taking fire when thrown into water. Chlorin in the free state is a poisonous gas. Thus the change in properties is well illustrated when these two dangerous elements, sodium and chlorin, unite to form the harmless compound which we call common salt.

"It is a shame," continued Percy, "that agricultural science has so long been burdened with such a term as 'phosphoric acid,' which serves to complicate and confuse what should be made the simplest subject to every American farmer and landowner. As agriculture is the fundamental support of America and of all her other great industries, so the fertility of the soil is the absolute support of every form of agriculture. Now, if there is any one factor that can be the most important, where so many are positively essential, then the most important factor in the problem of adopting and maintaining permanent systems of profitable agriculture on American soils is the element phosphorus.

"Phosphorus in very appreciable amount is positively necessary for the growth of every organism. It is an absolutely essential constituent of the nucleus of every living cell, whether plant or animal. Nuclein, itself, which is the substance nearest to the beginning of a new cell, contains as high as ten per cent. of the element phosphorus.

"On the other hand, phosphorus is the most limited of all the plant food elements, measured by supply and demand and circulation.

"What is phosphoric acid? Well, the professor of chemistry says it is a compound containing three atoms of hydrogen, one of phosphorus, and four of oxygen. It is a syrupy liquid and one of the strongest mineral acids. In concentrated form it is as caustic as oil of vitriol. Why, here you have a Century dictionary. That should tell what phosphoric acid is. This is what the Century says:

"'It is a colorless, odorless syrup, with an intensely sour taste. It is tribasic, forming three distinct classes of metallic salts. The three atoms of hydrogen may in like manner be replaced by alcohol radicles, forming acid and neutral ethers. Phosphoric acid is used in medicine as a tonic.'

"That," continued Percy, "is the complete definition as given by the Century dictionary as to what phosphoric acid is, and I note that this is the latest edition of the Century, copyrighted in 1902."

"We bought it less than a month ago," said Mrs. Thornton. "We can have so few books that we thought the Century would be a pretty good library in itself; Mr. Thornton has had too little time to use it much as yet."

"Well, even if I had used it," said Mr. Thornton, "you see there are five volumes before I'd get to the P's. But, joking aside, I don't get much out of that definition except that phosphoric acid is a sour liquid and is used in medicine."

"The definition is entirely correct," said Percy "Any text on chemistry will give you a very similar definition, and your physician and druggist will give you the same information."

"Well, I know the fertilizer agents claim to sell phosphoric acid in two-hundred-pound bags which wouldn't hold any kind of liquid."

"True," replied Percy, "and I consider it a shame that the farm boy who goes to the high school or college and is there taught exactly what phosphoric acid is, must. when he returns to the farm, try to read bulletins from his agricultural experiment station in which the term 'phosphoric acid' is used for what it is not. At the state agricultural college, the professor of chemistry correctly teaches the farm boy that phosphoric acid is a liquid compound containing three atoms of hydrogen, one of phosphorus, and four of oxygen in the molecule; and then the same professor, as an experiment station investigator, goes to the farmers' institutes and incorrectly teaches the same boy's father that phosphoric acid is a solid compound pound containing two atoms of phosphorus and five atoms of oxygen in the molecule."

"But why do they continue to teach such confusion?"

"Well, Sir, if they know, they never tell. In some manner this misuse of the name was begun, and every year doubles the difficulty of stopping it."

"Like the man that was too lazy to stop work when he had once begun," remarked Mr. Thornton.

"Yes," said Percy, "but it is true that some of the States have adopted the practice of reporting analyses of soils and fertilizers on the basis of nitrogen instead of ammonia; and in the Corn Belt States, phosphorus and potassium are the terms used to a large extent instead of 'phosphoric acid,' and potash. The agricultural press is greatly assisting in bringing about the adoption of the simpler system, and the laws of some States now require that the percentages of the actual plant food elements, as nitrogen, phosphorus, and potassium, shall be guaranteed in fertilizers offered for sale. It is one of those questions that are never settled until they are settled right; and it is only a question of time until the simple element basis will be used throughout the United States, or at least in the Central and Western States."

"The so-called 'phosphoric acid' of the fertilizer agent is a compound whose molecule contains two atoms of phosphorus and five atoms of oxygen; and, since the atomic weight of phosphorus is thirty-one and that of oxygen sixteen, this compound contains sixty-two parts of phosphorus and eighty parts of oxygen. In other words, this phosphoric acid, falsely so-called, contains a trifle less than forty-four per cent. of the actual element phosphorus."

"Is the bone phosphate of lime that the agents talk about the same as the 'phosphoric acid'?" asked Mr. Thornton.

"No, by 'bone phosphate of lime,' which is often abbreviated B. P. L., is meant tricalcium phosphate, a compound which contains exactly twenty per cent. of phosphorus. Thus, you can always divide the guaranteed percentage of 'bone phosphate of lime' by five, and the result will be the per cent. of phosphorus.

"As stated in your Century dictionary, true phosphoric acid forms three distinct classes of salts, because either one, two, or all of the three hydrogen atoms may be replaced by a metallic element. Thus, we have phosphoric acid itself containing the three hydrogen atoms, one phosphorus atom, and four oxygen atoms. This might be called trihydrogen phosphate ($H3PO4$). Now if one of the hydrogen atoms is replaced by one potassium atom, we have potassium dihydrogen phosphate ($KH2PO4$); with two potassium atoms and one hydrogen, we have dipotassium hydrogen phosphate ($K2HPO4$); and if all hydrogen is replaced by potassium the compound is tripotassium phosphate ($K3PO4$). To make similar salts with two-handed metallic elements, like calcium or magnesium, we need to start with two molecules of phosphoric acid $H6(PO4)2$; because each atom of calcium will replace two hydrogen atoms. Thus we have mono calcium phosphate, $CaH4(PO4)2$, dicalcium phosphate, $Ca2H2(PO4)2$, and tricalcium phosphate, $Ca_3(PO_4)_2$. It goes without saying that monocalcium phosphate contains four atoms of hydrogen and that dicalcium phosphate contains two hydrogen atoms. By knowing the atomic weights (40 for calcium, 31 for phosphorus, and 16 for oxygen), it is easy to compute that the molecule of tricalcium phosphate weighs 310 of which 62 is phosphorus. This is exactly one-fifth, or twenty per cent. This compound you will remember is sometimes called 'bone phosphate of lime'. It is also called simply 'bone phosphate'; because it is the phosphorus compound contained in bones. It is sometimes called lime phosphate, although it contains no lime in the true sense, for it has no power to neutralize acid soils, except when the phosphorus is taken up by plants more rapidly than the calcium, which in such case might remain in the soil to act as a base to neutralize soil acids; but even then the effect of the small amount of calcium thus liberated from the phosphate would be very insignificant compared with a liberal application of ground limestone."

"Well," said Mr. Thornton, stretching himself, "orange phosphate is my favorite drink but I fear some of these phosphate you have just been giving me are too concentrated. I ought to have the dose diluted; but I like the taste of it, and if you'll write a book along this line, in this plain way just about as you have been giving it to me straight for almost twelve hours, I tell you I'll read it over till I learn to understand it a heap better than I do now."

CHAPTER XVIII

CLOSER TO MOTHER EARTH

THE following day Percy collected soil samples to represent the common type of soil on the farm. In the main the land was nearly level and very uniform, although here and there were small areas which varied from the main type, and in places the variation was marked. Percy and his host devoted the entire day to an examination of the soils of the farm and the collection of the samples.

"The prevailing soil type is what would be called a loam," said Percy, "and a single set of composite samples will fairly represent at least three-fourths of the land on this farm.

"It seems to me that it is enough for the present to sample this prevailing type, and later, if you desire, you could collect samples of the minor types, of which there are at least three that are quite distinct."

"A loam soil is one that includes a fair proportion of the several groups of soil materials, including silt, clay, and sand."

"What is silt?" asked Mr. Thornton.

"Silt consists of the soil particles which are finer than sand,—too small in fact to be felt as soil grains by rubbing between the fingers, and yet it is distinctly granular, while clay is a mere plastic or sticky mass like dough. What are commonly called clay soils consist largely of silt, but contain enough true clay to bind the silt into a stiff mass. In the main such soils are silt loams, but when deficient in organic matter they are yellow in color as a rule, and all such material is usually called clay by the farmers."

"Well, I had no idea that it would take us a whole day to get enough dirt for an analysis," remarked Mr. Thornton, as they were collecting the samples late in the afternoon. "Five minutes would have been plenty of time for me, before I saw the holes you've bored to-day."

"The fact is," replied Percy, "that the most difficult work of the soil investigator is to collect the samples. Of course any one could fill these little bags with soil in five minutes, but the question is, what would the soil represent? It may represent little more than the hole it came out of, as would be the case where the soil had been disturbed by burrowing animals, or modified by surface accumulations, as where a stack may sometime have been burned. In the one case the subsoil may have been brought up and mixed with the surface, and in the other the mineral constituents taken from forty acres in a crop of clover may have been returned to one-tenth of an acre."

"Certainly such things have occurred on many farms," agreed Mr. Thornton, "and they may have occurred on this farm for all any one knows."

"Fifty tons of clover hay," continued Percy, after making a few computations, "would contain 400 pounds of phosphorus, 2400 pounds of potassium, 620 pounds of magnesium, and 2340 pounds of calcium."

"I don't see how you keep all those figures in your head," said Mr. Johnston.

"How many pounds are there in a ton of hay?" asked Percy.

"Two thousand."

"How many pounds in a bushel of oats?"

"Thirty in Virginia, but thirty-two in Carolina."

"How many in a bushel of wheat?"

"Sixty"

"Corn?"

"Fifty-six pounds of shelled corn, or seventy pounds of ears."

"Potatoes?"

"Eighty-six pounds,—both kinds the same, but most States require sixty pounds for the Irish potatoes."

Percy laughed. "You see," he said, "you have more figures in your head than I have in mine. You have mentioned twice as many right here, without a moment's hesitation, as I try to remember for the plant food contained in clover. I like to keep in mind the requirements of large crops, such as it is possible to raise under our climatic conditions if we will provide the stuff the crops are made of, so far as we need to, and do the farm work as it should be done. I never try to remember how much plant food is required for twenty-two bushels of corn per acre, which is the average yield of Virginia for the last ten years, while an authentic record reports a yield of 239 bushels from an acre of land in South Carolina. On our little farm in Illinois we have one field of sixteen acres, which was used for a pasture and feed lot for many years by my grandfather and has been thoroughly tile-drained since I was born, that has produced as high as 2,015 bushels of corn in one season, thus making an average of 126 bushels per acre.

"What I try to remember is the plant food requirements for such crops as we ought to try to raise, if we do what ought to be done. I try to remember the plant food required for a hundred-bushel crop of corn, a hundred-bushel crop of oats, a fifty-bushel crop of wheat, and four tons of clover hay. It is an easy matter to divide these amounts by two, as I have really been doing here in the East where it is hard for people to think in terms of such crops as these lands ought to be made to produce.

"The requirements of the clover crop I certainly want to have in mind as a part of my little stock of ever-ready knowledge. It is not very hard to remember that a four-ton crop of clover hay, which we ought to harvest from one acre in two cuttings, contains:

160 pounds of nitrogen, 31 pounds of magnesium, 20 pounds of phosphorus, 120 pounds of potassium, 117 pounds of calcium.

"It is just as easy to think in these terms as in per cent. or pounds of butter fat, which I understand is the basis on which you sell your cream."

"Yes, I believe you are right in this matter, Mr. Johnston, but I have never been able to see how we could apply the figures reported from chemical analysis."

"Neither do I see how any one but a chemist could make much use of the reports which the analyst usually publishes. Such reports will usually show the percentages of moisture and so-called 'phosphoric acid,' for example, in a sample of clover hay, and perhaps the percentages of these constituents in a sample of soil; but to connect the requirements of the clover crop with the invoice of the soil demand more of a mental effort than I was prepared for before I went to the agricultural college.

"On the other hand we were taught in college that the plowed soil of an acre of our most common Illinois corn belt land contains only 1200 pounds of phosphorus, and that a hundred-bushel crop of corn takes twenty-three pounds of phosphorus out of the soil. Furthermore that about one pound of phosphorus per acre is lost annually in drainage water in humid regions. By dividing 1200 by 24 it is easy to see that fifty corn crops such as we ought to try to raise would require as much phosphorus as the present supply in our soil to a depth of about seven inches. Of course there is some phosphorus below seven inches, but it is the plowed soil we must depend upon to a very large extent. The oldest agricultural experiment station in the world is at Rothamsted, England. On two plots of ground in the same field where wheat has been grown every year for sixty years, the soil below the plow line has practically the same composition, but on one plot the average yield for the last fifty years has been thirteen bushels per acre, while on the other the yield of wheat has averaged thirty-seven bushels for the same fifty years."

"The same kind of wheat?" inquired Mr. Thornton.

"Yes, and great care has always been taken to have these two plots treated alike in all respects, save one."

"And what was that?"

"Plant food was regularly incorporated with the plowed soil of the high-yielding plot."

"You mean that farm manure was used?"

"No, not a pound of farm manure has been used on that plot for more than sixty years; and, furthermore, the two plots were very much alike at the beginning; but, to the high-yielding plot, nitrogen, phosphorus, potassium, magnesium, calcium, and sulfur have all been applied in suitable compounds every year."

"That is to say," observed Mr. Thornton, "that the land itself has produced thirteen bushels of wheat per acre and the plant food applied has produced twenty four bushels, making the total yield thirty-seven bushels on the fertilized land."

"That is certainly a fair way to state it," replied Percy.

" Well, that sounds as though something might be done with run-down lands. About what part of the twenty-four bushels increase would it take to pay for the fertilizers?"

"About 150 per cent. of it," Percy replied.

"One hundred and fifty per cent! Why, you can't have more than a hundred per cent. of anything."

"Oh, yes, you can. The twenty-four bushels are one hundred per cent. of what the fertilizers produced, and the land itself increased this by fifty per cent., so that the fertilized land produced one hundred and fifty per cent. of the increase from the plant food applied.

"Well, that's too much college mathematics for me; but do you mean to say that it would take the whole thirty-seven bushels to pay for the plant food that produced the increase of twenty-four bushels?"

"That is exactly what I mean. I see that you do not like percentage any better than I do. Really the acre is the best agricultural unit. We buy and sell the land itself by the acre; we report crop yields at so many bushels or tons per acre; we apply manure at so many loads or tons per acre; we apply so many hundred pounds of fertilizer per acre; sow our wheat and oats at so many pecks or bushels per acre; and we ought to know the invoice of plant food in the plowed soil of an acre and the amounts carried off in the crops removed from an acre.

"Now, referring again to these figures from the forty acres of clover at two tons per acre. If the eighty tons were burned and the ashes mixed with the surface soil on a tenth of an acre the increase per acre would be as follows:

4,000 pounds of phosphorus 24,000 pounds of potassium 6,200 pounds of magnesium 23,400 pounds of calcium.

"These, remember, are the amounts per acre that would be added to the soil by burning the eighty tons of clover on one-tenth of an acre.

"Now compare these figures with the total amounts of the same elements contained in the common corn belt prairie soil of Illinois, which are as follows:

1,200 pounds of phosphorus 35,000 pounds of potassium 8,600 pounds of magnesium 5,400 pounds of calcium.

"From these figures you will see that the analysis of a single sample of soil collected from a spot of ground that had sometimes received such an addition as this would be positively worse than worthless, because it would give false information, and that is much worse than no information.

"The methods of chemical analysis have been developed to a high degree of accuracy, and it is not a difficult matter to find a chemist who can make a correct analysis of the sample placed in his hands; but the chief difficulties lie, first, in securing samples of soil that will truly represent the type or types of soil on the farm; and, second, in the interpretation of the results of analysis with reference to the adoption of methods of soil improvement."

"Is the report of the analysis as confusing with respect to other elements as with potassium and phosphorus, which, I understand, are likely to be reported in terms of potash and a 'phosphoric acid' that is not true phosphoric acid?"

"Still worse," Percy replied. "The calcium is commonly reported in terms of lime, or, as you would say, quick lime; and vet the soil may be an acid soil, like yours, and contain no lime whatever, neither as quick lime nor limestone. I have seen an analysis reporting half a per cent. of calcium oxid, which would make five tons of quick lime in the plowed soil of an acre; whereas the soil not only contained no lime whatever, but was so acid that it needed five tons of ground limestone per acre to correct the acidity.

"The trouble is that when the chemist found calcium in the soil existing in the form of acid silicate, or calcium hydrogen silicate, he reported calcium oxid, or lime, in his analytical statement, assuming apparently that the farmer would understand that the analytical statement did not mean what it said."

"But some soils do contain lime, do they not?"

"Some soils contain limestone," replied Percy, "and the analysis of such a soil should report the amount of limestone, or calcium carbonate, based upon the actual determination of carbonate carbon or carbon dioxid, which is a true measure of the basic property of the soil, even though the limestone may be somewhat magnesian in character."

For a set of soil samples. Percy collected soil from three different strata. The first sample represented the surface stratum from the top to six and two-third inches; the second sample represented the subsurface stratum from six and two-thirds to twenty inches; and the third sample represented the subsoil from twenty to forty inches, each sample being a composite of about twenty borings.

In collecting these the hole was bored to six and two-third inches and somewhat enlarged by scraping up and down with the auger, all of the soil being put into a numbered bag. Then, the hole was extended and the subsurface boring removed without touching the surface soil. This boring to a depth of twenty inches was put into a second bag. The hole was then enlarged to the twenty-inch depth but the additional soil removed was discarded as a mixture of the surface and subsurface strata. Finally the hole was extended to the forty-inch depth and the subsoil from one groove of the auger was put into a third bag. In this manner about an equal quantity of soil was bagged from each stratum; and twenty such borings taken with an auger about one inch in diameter make a sufficient quantity to furnish to the chemist.

"Of course the surface soil is by far the most important," Percy explained. "It represents just about the depth of earth that is turned by the plow in good farming on normal soils; and it weighs about two million pounds per acre. The subsurface stratum extending from six and two-thirds to twenty inches in depth represents the practical limit of subsoiling; and this stratum weighs about four million pounds; while the subsoil stratum weighs about six million pounds, where the soil is normal, such as loam, silt loam, clay loam, or sandy loam. Pure sand soil weighs about one-fourth more, while pure peat soil weighs only half as much as normal soil."

"I wish you would tell me," said Mr. Thornton, "what the fertilizers cost that have been used on that Rothamsted wheat field."

"The annual application of nitrogen has been one hundred twenty-nine pounds per acre," said Percy. "What will it cost?"

"Well, at twenty cents a pound, it would cost $25.80," was Mr. Thornton's reply after he had figured a moment. "But why didn't they grow clover and get the nitrogen from the air?"

"For two reasons," replied Percy. "First, when those classic experiments were begun by Sir John Lawes and Sir Henry Gilbert in 1844, it was not known that clover could secure the free nitrogen from the air; and, second, the experiment was designed to discover for certain whether wheat must be supplied with combined nitrogen, by ascertaining the actual effect upon the yield of wheat of the nitrogen applied."

"And what was the actual effect of the nitrogen?" questioned Mr. Thornton. "How much did the wheat yield when they left out the nitrogen and applied all the other elements?"

"Only fifteen bushels," was the reply.

"Only fifteen bushels! Only two bushels increase for all the other elements, phosphorus, potassium, magnesium, and calcium,—and I remember you said that sulfur also was applied. Why didn't they leave off all these other elements, and just use the nitrogen alone?"

"They did on another plot in the same field."

"Oh, they did do that? What was the yield on that plot?"

"Only twenty bushels."

"Only twenty bushels! Well, that s mighty queer. How do you account for that?"

"Does Mrs. Thornton sometimes make dough out of flour and milk?" asked Percy.

"Another Yankee question, eh?" said Mr. Thornton. "I told my wife once that I wished she could make the bread my mother used to make, and she said she wished I could make the dough her father used to make. Yes, my wife makes dough, a good deal more than I do, and she makes it of flour and milk, when we aren't reduced to corn meal and water."

"Can she make dough of flour alone?" continued Percy.

"No," replied Mr. Thornton.

"Nor of milk alone?"

"No."

"Well, wheat cannot be made of nitrogen alone, nor can it be made without nitrogen. On Broadbalk field at Rothamsted, where the wheat is grown, the soil is most deficient in the element nitrogen. In other words, nitrogen is the limiting element for wheat on that soil; and practically no increase can be made in the yield of wheat unless nitrogen is added. However, some other elements are not furnished by this soil in sufficient amount for the largest yield of wheat, and these place their limitation upon the crop at twenty bushels. To remove this second limitation requires that another element, such as phosphorus, shall be supplied in larger amount than is anually liberated in the soil under the system of farming practiced."

"Yes, I see that," said Mr. Thornton, "it's like eating pancakes and honey; the more cakes you have the more honey you want. I think I can almost see my way through in this matter; we are to correct the acid with limestone, to work the legumes for nitrogen, and turn under everything we can to increase the organic matter, and if we find that the soil won't furnish enough phosphorus, potassium, magnesium, or calcium, even with the help of the decaying organic matter to liberate them, why then it is up to us to increase the supply of those elements."

"You must remember that the calcium will be supplied in the limestone;" cautioned Percy. "And, if you use magnesian limestone, you will thus supply both calcium and magnesium. Keep in mind that _magnesian _only means that the limestone contains some *magnesium.* and that it is not a pure calcium carbonate. The purest magnesian limestone consists of a double carbonate of calcium and magnesium, called dolomite."

"But I have heard that magnesian lime is bad for soils," said Mr. Thornton.

"That is true," Percy replied, "and so is ordinary lime bad for soils. The Germans say: 'Lime makes the fathers rich but the children poor.' The English saying is:

'Lime and lime without manure
Will make both farm and farmer poor.'

"Both of these national proverbs are correct for common, everyday lime; but you know, do you not, that limestone soils are usually very good and very durable soils?"

"That's what I've always heard," replied Mr. Thornton.

"Well, there is no danger whatever from using too much limestone; and all the information thus far secured shows that magnesian limestone is even better than the pure calcium limestone. I know two Illinois farmers who are using large quantities of ground magnesian limestone, and one of them has applied as much as twenty tons per acre. On that land his corn crop was good for eighty bushels per acre this year. Of course that heavy application was more than was needed, but initial applications of four or five tons are very satisfactory, and these should be followed by about two tons per acre every four to six years."

Mr. Thornton took his guest to Blairville that evening as they had planned and he assured Percy that should he decide to purchase land in that section they would let him have three hundred acres of their land at ten dollars an acre.

"I will let you know after I get the samples analyzed for you," said Percy. "The price is low enough and the location ideal, but still I want to have the invoice before I buy the goods. I will write you about sending the samples to the chemist after I hear from some I sent him from Montplain."

CHAPTER XIX

FROM RICHMOND TO WASHINGTON

THE next day Percy spent a few hours at the State Capitol in Richmond, where he found the records of the State of much interest.

Thus he found that in practically every county there was more or less land owned by the commonwealth, because of its complete abandonment by former owners, and the failure of any one to buy when sold by the state for taxes.

Under such conditions the title to the land returns to the State, and after two years it may be sold by the State to any one desiring to purchase and the former owner has no further right of redemption. Some of these lands which are owned by the State, and on which the State has received no taxes for many years, are still occupied by their former owners or by "squatters"' and may continue to be so occupied unless the land should be purchased from the State by some one else who would demand full possession. Such purchasers, however, are likely to be unpopular residents in the community, if the transaction forces poor people from a place they have called home, even though they had no legal right to occupy it.

Percy found that the report of the State Auditor showed that the clerk of the court of Powhatan county had returned to the State $1.05 "for sales of lands purchased by the commonwealth at tax sales," while from Prince Edward county the State received a similar revenue amounting to $17.39 for the same year. The total revenue to the commonwealth from this source amounted to $667.85 for the year. Contrasted with this was the revenue from "Redemption of Land," amounting to $27,436.38, suggesting something of the struggle of the man to retain possession of his home before it becomes legally possible for another to take it from him beyond redemption.

According to the records about a million acres of land are owned by the Commonwealth of Virginia alone.

Percy decided to go to Washington to learn what definite information he might obtain from the United States Department of Agriculture. On the train for Washington he found himself sitting beside a Virginia farmer.

"These lands remind me of our Western prairies," Percy remarked. "You have some extensive areas of level or gently undulating uplands."

"They don't remind me of the Western prairies, I can tell you," was the reply. "I am a Westerner myself, or I was until eight years ago. These lands look all right from the train when the crops are all off, but I find that every patch of the earth's surface doesn't always make a good farm. Why you can go from Danville, Illinois, to Omaha, Nebraska, and stop anywhere in the darkest night and you're mighty near sure to light on a good farm where one acre is worth ten of this land along here."

"About what is this land worth?" asked Percy.

"Well, I thought six hundred acres of it was worth $5,000 about eight years ago, especially as the buildings on the place were in good repair and couldn't be built to-day for less than $6,000: but right now I think I paid a plenty for my land. It's just back a few miles at the station where I got on."

"How far is that from Washington?"

"About fifteen miles, I reckon, as the crow flies. My boy has a telescope his uncle sent him and we can see the Monument on a clear day."

"What monument?" asked Percy.

"Why, Washington's monument. Haven't you ever been to Washington?"

"No, this is my first visit. I am really thinking of buying a farm somewhere here in the East. I have been in Richmond and learned a great deal from the state reports, and I thought I might get more information from the Department of Agriculture in Washington."

"Perhaps," said the man, "but my advice is to keep in mind that there is a difference between buying land and buying a farm. I've got land to sell, by the way. I thought I'd need it all when I bought, but I can see now that I'll not need more'n half of it at the most; so, if you want two or three hundred acres of this kind of land right close here where you kind o' neighbor with the senators and other upper tens, and run back and forth from the City in an hour or so, why I think I can accommodate you. My name is Sunderland, J. R. Sunderland, and you'll find me at home any day."

"How much would you sell part of your land for?" inquired Percy.

"Well, I'd kind o' hate to take less than ten dollars an acre for it; but I think we can make a deal all right if you like the location."

CHAPTER XX

A LESSON IN OPTIMISM

ABOUT nine o'clock the day following Percy's arrival in Washington he sent his card into the office of the Secretary of Agriculture.

"Just step this way," said the boy on his return. "The Secretary will see you at once."

A gentleman who appeared to be sixty, but was really several years older, arose from his desk and greeted Percy very kindly.

"I see you are from Illinois, Mr. Johnston. I am an Iowa man myself, and I am always glad to see any one from the corn belt. Do you know we are going to beat the records this year? It is wonderful what crops we grow in this country, and they are getting better every year. We are growing more than two-thirds of the entire corn crop of the globe, right here in these United States. Yes, Sir, and we are just beginning to grow corn; and corn is only one of our important agricultural products. Do you know that eighty-six per cent. of all the raw materials used in all the manufactured products of this country come from the farms of the United States; yes, Sir, eighty-six per cent.

"Now, what can I do for you? I am very glad you called, and I will be glad to serve you in any way you desire. By the way, how is the corn turning out in your part of Illinois? Bumper crop, I have no doubt."

"I think so," said Percy, "after seeing the crops here in the East.

"That's what I thought," continued the Secretary." A bumper crop, the biggest we ever raised. Oh, they don't know how to raise corn here in the East. They just grow corn, corn, corn, year after year; and that will get any land out of fix. I found that out years ago in Iowa. I am a farmer myself, as I suppose you know. I found you couldn't grow corn on the same land all the time. But just rotate the crops; put clover in the rotation; and then your ground will make corn again, as good as ever."

"But I understand that clover refuses to grow on most of this eastern land," said Percy.

"Oh, nonsense. They don't sow it. I tell you they don't sow it, and they don't know how to raise it. It takes a little manure sometimes to start it, but it will grow all right if they would only give it half a chance. Why, for years the Iowa farmers said blue grass wouldn't grow in Iowa. Yes, Sir, they just knew it wouldn't grow there; and then I showed them that blue grass was actually growing in Iowa,—actually growing along the roadsides almost everywhere,—blue grass that would pasture a steer to the acre—just came in of itself without being seeded. No, I tell you they don't sow clover down here. They just say it won't grow and keep right on planting corn, corn, corn, until the corn crop amounts to nothing, and then they let the land grow up in brush."

"Now, I do not wish to take up more of your time," said Percy, "for I know how busy a man you must be, but I am thinking of buying a farm, or some land, here in the East and have come to you for information. We have a small farm in Illinois and land is rather too high-priced there to think of buying more; but I thought I could sell at a good price, and buy a much larger farm here in the East with part of the money and still have enough left to build it up with; and, with the high price of all kinds of farm produce here, we ought to make it pay."

"You can do it," said the Secretary. "No doubt of it. Any land that ever was any good is all right yet if you'll grow clover, and you can start that with a little manure if you need it. I have done it in Iowa, and I know what I am talking about.

"Now my Bureau of Soils can give you just the information you want. We are making a soil survey of the United States, and we have soil maps of several counties right here in Maryland. You can take that map and pick out any kind of land you want,—upland or bottom land,—sandy soil, clay soil, loam, silt loam, or anything you want."

CHAPTER XXI

IN THE OFFICE OF THE CHIEF

"SHOW this gentleman to the Bureau of Soils," said the Secretary to the boy who came as he pushed a button.

"All the world loves an optimist," said Percy to himself as he followed the boy to another office where he met the Chief of the Bureau of Soils, who kindly furnished him with copies of the soil maps of several counties, including two in Maryland, Prince George, which adjoins the District of Columbia, and St. Mary county, which almost adjoins Prince George on the South.

These maps were accompanied by extensive reports describing in some detail the agricultural history of the counties and the general observations that had been made by the soil surveyors.

"I desire to learn as much as I can regarding the most common upland soils," Percy explained. "Not the rough or broken land, but the level or undulating lands which are best suited for cultivation. I am sure these maps and reports will be a very great help to me."

"I think you will find just what you are looking for," said the Chief. "You can spread the maps out on the table there and let me know if I can be of any assistance. You see the legend on the margin gives you the name of every soil type, and the soils are fully described in the reports. One of the most common uplands soils in southern Prince George county is the Leonardtown loam, and this type is also the most extensive soil type in St. Mary county.

"The same type is found in Virginia to some extent. While the soil has been run down by improper methods of culture, it has a very good mechanical composition and is really an excellent soil; but it needs crop rotation and more thorough cultivation to bring it back into a high state of fertility. The farmers are slow to take up advanced methods here in the East. We have told them what they ought to do, but they keep right on in the same old rut."

For two hours Percy buried himself with the maps and reports. Finally the Chief came from his inner office, and finding Percy still there asked if he had found such information as he desired.

"I find much of interest and value, but I do not find any complete invoice of the plant food contained in these different kinds of soil."

"You mean an ultimate chemical analysis of the soil?" asked the Chief.

"Yes, a chemical analysis to ascertain the absolute amount of plant food in the soil. I think of it as an invoice; but I see that you do not report any such analyses."

"No, we do not," answered the Chief. "We have been investigating the mechanical composition of soils, the chemistry of the soil solution, and the adaptation of crop to soil. We find that farmers are not growing the crops they should grow; namely, the crops to which their soils are best adapted. For example, they try to grow corn on land that is not adapted to corn."

"It seems to me," said Percy, "that our farmers are always trying to find a crop that is adapted to their soil. Down in 'Egypt,' which covers about one-third of Illinois, the farmers once raised so much corn that the people from the swampy prairie went down there to buy corn, and hence the name 'Egypt' became applied to Southern Illinois. But there came a time when the soil refused to grow such crops of corn; the farmers then found that wheat was adapted to the soil. Later the wheat yields decreased until the crop became unprofitable; and the farmers sought for another crop adapted to a still more depleted soil. Timothy was selected, and for many years it proved a profitable crop; but of late years timothy likewise has decreased in yield until there must be another change; and now whole sections of 'Egypt' are growing red top as the only profitable crop. After red top, then what? I don't know, but it looks as though it would be sprouts and scrub brush, and final land abandonment, a repetition of the history of these old lands of Virginia and Maryland."

"Well, can't they grow corn after red top?" asked the Chief.

"Many of them try it many times," replied Percy, "and the yield is about twenty bushels per acre, whereas the virgin soil easily produced sixty to eighty bushels."

"And they can't grow wheat as they once did?"

"No, wheat after timothy or red top now yields from five to twelve bushels per acre, while they once grew twenty to thirty bushels of wheat per acre year after year.

"If they rotate their crops, they would probably yield as well as ever," said the Chief.

"No, that, too, has been tried," replied Percy. "The Illinois Experiment Station has practiced a four-year rotation of corn, cowpeas, wheat, and clover on an experiment field on the common prairie soil down in 'Egypt,' and the average yield of wheat has been only twelve bushels per acre during the last four years, but when legume crops were plowed under and limestone and phosphorus applied, the average yield during the same four years was twenty-seven bushels per acre."

"Probably the increase was all produced by the green manure," suggested the Chief. "Organic matter has a great influence on the control of the moisture supply."

"That was tested," said Percy. "The green manure alone increased the average yield to only fourteen bushels while the green manure and limestone together raised the average wheat yield to nineteen bushels, the further increase to twenty-seven bushels having been produced by the addition of phosphorus."

"Well, Sir," said the Chief, "we have made both extensive intensive investigations concerning the chemistry of the soil solution by very delicate and sensitive methods of analysis we have developed, and we have also conducted culture experiments for twenty-day periods with wheat seedlings in the water extract of soils from all parts of the United States, and the results we have obtained have changed the thought of the world as to the cause of the infertility of soils."

"But you have not made analyses for total plant food in the soils or conducted actual field experiments with crops grown to maturity?" asked Percy.

"No, we have not done that," answered the Chief. "Those are old methods of investigation which have been tried for many years and yet no chemist can tell in advance what will be the effect of a given fertilizer upon a given crop on a given soil."

"That is true," said Percy, "but neither can any merchant tell in advance just what effect will be produced on the next day's business by the addition of a given number of a given kind of shoes to a given stock on his shelves. There are many factors involved in both cases."

"Yes, you are right in that," said the Chief, "we are just beginning to understand the chemistry of the soil, and we hope soon to have very complete proof of the advanced ideas we already have concerning the causes of the fertility and infertility of soils."

"Referring to the specific case of the Leonardtown loam of Maryland," said Percy, "I find the following statement on page 33 of the Report of the Field Operations of the Bureau of Soils for 1900. After describing the Norfolk loam of St. Mary County, the writer says:

"'The Leonardtown loam is a very much heavier type of soil. It covers about forty-one per cent. of St. Mary County. The soil is a yellow silty soil, resembling loess in texture, underlaid by a clay subsoil with layers or pockets of sand. This soil has been cultivated for upward of two hundred years, but it is now little valued and is covered with oak and pine over much of its area. It is worth from $1 to $3 per acre. The cultivated areas produce small crops of corn, wheat, and an inferior grade of tobacco.'"

"The generally low estimation in which this land is held is probably wholly unjustified," replied the Chief. "There are two or three farms in the area which, under a high state of cultivation with intelligent methods, will produce from twenty to thirty bushels of wheat per acre and corresponding crops of corn. Those farmers are a credit to the country. They furnish the towns with good milk and butter and vegetables, and they also help to keep the towns clean and sanitary by hauling out the animal excrements, and other waste and garbage that tend to pollute the air and water of the village."

"I can see how that might maintain the fertility of those farms," said Percy. "It seems that the general condition of this kind of land is about the same in Prince George County. On page 45 of the 1901 Report of the Field Operations of the Bureau of Soils, I have noted the following statement:

"'The Leonardtown loam, covering 45,770 acres of the area, is the nearest approach among the Maryland Coastal Plain Soils to the heavy clays of the limestone regions of Western Maryland and Pennsylvania. The surface is generally level and the drainage fair. The soil is not adapted to tobacco, and has consequently been allowed to grow up to scrub forest, so that large portions of it are at present uncleared. Such unimproved lands can be bought for $1.50 to $5.00 an acre, even within a few miles of the District line. The soil has been badly neglected, and when cultivated the methods have not been such as to promote fertility. When properly handled, as it is in a few places, good yields of wheat, corn and grass are obtained.'"

"That's right," said the Chief, "exactly right. Upon the whole it is one of the most promising soils of the locality, although it is not considered so by the resident farmers."

"You mean that it should be handled the same as is done by the successful farmers of St. Mary County?" inquired Percy.

"Yes, it needs thorough cultivation and the rotation of crops; and the physical condition of the soil needs to be improved by the addition of lime and manure, or green crops turned under."

"I have been looking over some of the other Reports of Field Operations," said Percy." I became interested in the description of a Virginia soil called Porters black loam. I find the following statements on page 210 of the Report for 1902:

"'The Porters black loam occurs in all the soil survey sheets, extending along the top of the main portion of the Blue Ridge Mountains in one continuous area. This type consists of the broad rolling tops and the upper slopes of the main range of the Blue Ridge Mountains. Locally the Porters black loam is called "black land" and "pippin" land, the latter term being applied because, of all the soils of the area, it is pre-eminently adapted to the Newtown and Albermarle Pippin. This black land has long been recognized as the most fertile of the mountain soils. It can be worked year after year without apparent impairment of its fertility. Wheat winter kills, the loose soils heaving badly under influence of frost. The areas lie at too high elevations for corn. Oats do well, making large yields. Irish potatoes, even under ordinary culture, will yield from two hundred to three hundred bushels per acre. It seeds in blue grass naturally, which affords excellent pasturage. Clover and other grasses will also grow luxuriantly upon it. The areas occupied by this soil are mostly cleared.'"

"Yes, Sir," said the Chief, "the Potters black loam is a fine soil—loose and porous as stated in the Report. You see it has a good physical condition."

"There is one other description in this Report for 1903 that is of special interest to me," said Percy. "This relates to a type of soil which the surveyors found in the low level areas of prairie land in McLean County, Illinois, and which they have called Miami black clay loam. I think we have several acres of the same kind of soil on our own little farm. I found the following statements on page 787:

"'When the first settlers came to McLean County they found the areas occupied by the Miami black clay loam wet and swampy, and before these areas could be brought under cultivation it was necessary to remove the excess of moisture. With the exception of a few large ditches for outlets, tile drains have taken the place of open ditches. Drainage systems in some instances have cost as much as $25 an acre, but the very productive character of the soil, and the increase in the yields fully justify the expense. There are few soils more productive than the Miami black clay loam. Some areas have been cropped almost continuously in corn for nearly fifty years without much diminution in the yields.'"

"Now there you are again," said the Chief. "Drainage, that's all it needed. You see it's a simple matter; and that's what the Leonardtown loam needs in places. Give it good drainage and good cultivation with a rotation of crops, and you'll get results all right."

"Has the Bureau of Soils tried these methods on any of this soil near Washington?" asked Percy.

"No use," replied the Chief. "We've got the scientific facts and besides, as I told you, some few farms are kept up in both Prince George and St. Mary counties and they are as good demonstrations as anyone could want. Now I suggest that you meet some of our scientists."

CHAPTER XXII

THE CHEMIST'S LABORATORY

THE Chief showed Percy into the laboratories of the Bureau and introduced him to the soil physicist and the soil chemist. Percy was greatly interested in the various lines of work in progress and gladly accepted an invitation to return after lunch and become better acquainted with the methods of investigation used.

In the afternoon the physicist showed him how the soil water could be removed from an ordinary moist soil by centrifugal force, and the chemist was growing wheat seedlings in small quantities of this water and in water extracts contained in bottles. The seedlings were allowed to grow for twenty days and then other seedlings were started in the same solution and also in fresh solution, and it was very apparent that in some cases the wheat grew better in the fresh solutions.

The chemist explained that he also analyzed the soil solutions and water extracts from different soils and that there was no relation between the crop yields and the chemical composition of the soils.

"But it seems to me," said Percy, "that your analysis refers to the plant food dissolved in the soil water only at the time when you extract it. How long a time does it require to make the extraction?"

"As a rule we shake the soil with water for three minutes and then it takes twenty minutes to separate the water from the soil. This gives us the plant food in solution and with the addition of more water the nitrates, phosphoric acid, and potash in the soil immediately dissolve sufficiently give us a nutrient solution of the same concentration as we had before. Thus there is always sufficient plant food in the soil so long as there is any of the original stock."

"That is surely quick work," said Percy, "but I wonder if the corn plant might not get somewhat different results from the soil analysis which it makes."

"How do you mean?"

"Did you ever plant a field of corn and then cultivate it and watch it grow with increasing rapidity, until along about the Fourth of July every leaf seemed to nod its appreciation and thanks as you stirred the soil; and to show its gratitude, too, by growing about five inches every twenty-four hours when the nights were warm?"

"No," replied the Chemist, "I have never had any experience of that sort. I am devoting my life to the more scientific investigations relating to the fundamental laws which underlie these soil fertility problems."

"Well, I was only thinking," Percy continued, "that you analyze a fraction of a pound of soil in a few minutes, while the corn plant analyzes about a ton of soil by a sort of continuous process, which covers twenty-four hours every day for about one hundred and twenty days, and it takes into account every change in temperature and moisture, the aeration with any variation produced by cultivation, and also the changes brought about by the nitrifying bacteria and all other agencies that promote the decomposition of the soil and the liberation of plant food, including the action upon the insoluble phosphates and other minerals of the carbonic acid exhaled by the roots of the corn plants, the nitric acid produced by the process of nitrification, and the various acids resulting from the decay of organic matter contained in the soil."

"I am very familiar with the literature of the whole subject of soil fertility," replied the Chemist, "and our theories are being accepted everywhere. I have just returned from a lecture tour extending from Florida to Michigan, and our ideas and methods are being very generally adopted, not only in this country but also in Europe."

"The Chief of the Bureau very kindly permitted me to look over the maps and reports relating to the soils of Maryland and Virginia," said Percy, "but in this literature I found no data as to the amount of plant food contained in the various soil types that have been found in the surveys. May I ask if the Bureau has made any analyses to ascertain the total amounts of the different essential plant food elements contained in these different soils?"

"No," the Chemist replied, "a chemical analysis gives practically no information concerning the fertility of the soil. We have made no ultimate analyses of soils, although we have used the same methods of analysis in a study of the partial composition of the soil separates, or particles of different grades, such as the sand, the silt, and the clay."

"And have you also determined the percentages of sand, silt, and clay in the soils themselves?"

"Oh, yes, the physical composition of the soil is a matter of very great importance, and this is always determined and reported for every soil. Did you not see that in the Reports you examined this morning?"

"I think I did notice it," Percy replied, "but it is so easy for the farmer himself to tell a sandy soil from a clay soil that I did not appreciate the value of those physical analyses.

"In any case, I shall be very glad to know what results were obtained from the chemical analysis of the soil separates to which you referred."

"Those results are all reported in Bulletin No. 54 of the Bureau of Soils," said the Chemist, "and I have extra copies right here and will be glad to present you with one. And let me give you our Bulletin 22 also. This will enable you to get a clear idea of the principles we are developing which are solving the soil fertility problems that have completely baffled the scientists heretofore."

CHAPTER XXIII

MATHEMATICS APPLIED TO AGRICULTURE

PERCY left the Bureau of Soils with a feeling of deep appreciation for the uniform courtesy and kindness that had been accorded him, but with a firm conviction that the laboratory scientists were too far removed from the actual conditions existing in the cultivated field. He sought the quiet of his room at the hotel in order to study the bulletins he had received.

Even with his college training he found it difficult to form clear mental conceptions of the results of investigations reported in the bulletins. Sometimes the data were reported in percentages and sometimes in parts per million. No reports gave the amounts of the element phosphorus; but PO4 was given in some places and P2O5 in others. In Bulletin No. 22, the potassium and calcium were reported as the elements and the nitrogen in terms of NO3, while potash (K20), quicklime (CaO), and magnesia (MgO) were reported in Bulletin 54.

By a somewhat complicated mathematical process, he finally succeeded in making computations from the percentages of the various compounds reported in the soil separates and from the percentages of these different separates contained in the soils themselves and from the known weights of normal soils, until he reduced the data to amounts per acre of plowed soil.

He was especially pleased to find that the essential data were at hand not only for both the Leonardtown loam and the Porter's black loam, but also for the Norfolk loam, which he had learned from one of the soil maps was the principal type of soil southwest of Blairville on Mr. Thornton's farm; and, furthermore, the Miami black clay loam of Illinois was included. Percy knew the black clay loam was a rich soil, for the teacher in college had said that the more common prairie land and most timber lands were much less durable and needed thorough investigation at once, while the flat recently drained heavy black land could wait a few years if necessary.

Percy first worked out the data for the Miami black clay loam. The chemist had analyzed the soil separates for only four constituents, and they showed the following amounts per acre of plowed soil to a depth of six and two-thirds inches, averaging two million pounds in weight:

2,970 pounds of phosphorus

38,500 pounds of potassium

18,440 pounds of magnesium

46,200 pounds of calcium

He then made the computations for the average of the Leonardtown loam of St. Mary County, Maryland, with results as follows:

160 pounds of phosphorus

18,500 pounds of potassium

3,480 pounds of magnesium

1,000 pounds of calcium

Percy stared at these figures when he brought them together for comparison. He then checked up his computations to be sure they were right.

"Almost twenty times as much phosphorus!" he said to himself. "Is it possible? And more than forty times as much calcium! Let me see! It takes one hundred and seventeen pounds of calcium for four tons of clover hay. The total amount in the plowed soil of the Leonardtown loam would not be sufficient for eight such crops; and six crops of corn such as we raised one year on our sixteen acres would take more phosphorus from the land than is now left in the plowed soil of this Leonardtown loam. The magnesium is not quite so bad—about one-fifth as much as in our black soil, and the potassium is almost one-half as much as we have."

Percy next turned to the Porters black loam, which he had noticed was to be found not many miles from Montplain. He thought he might induce Mr. West to drive with him to the upper mountain slope in order that they might see that land. His computations for the Porters black loam gave the following results:

4,630 pounds of phosphorus

48,300 pounds of potassium

12,360 pounds of magnesium

23,700 pounds of calcium

He viewed these figures a moment with evident satisfaction.

"Plenty of everything in this wonderful 'pippin land,'" he thought. "Big yields reported for everything suited to that altitude. 'Can be worked year after year without apparent impairment of its fertility,' so the Report stated. I should think it might, especially since clover is one of the crops grown. Both phosphorus and potassium are way above our best black land. Magnesium two-thirds and calcium one-half of our flat land, but still greater than our common prairie, according to the average they gave us at college. And no doubt there is plenty of magnesian limestone in these mountains which could be had if ever needed. The soil surveyor certainly did not say too much in praise of the Porters black loam, considering that its physical composition is also all right."

He worked out the Norfolk loam to see what he would get if he accepted Miss Russell's dare. The following are the figures:

610 pounds of phosphorus

13,200 pounds of potassium

1,200 pounds of magnesium

3,430 pounds of calcium

"Rather low in everything," said Percy, "compared with any soil I know that has a good reputation. More uniformly poor but not so extremely poor as the Leonardtown loam."

He wished that the nitrogen had been determined by the chemist, even though he knew the organic matter and the nitrogen must be very low in the poor soils, but nowhere was any such record to be found in the bulletin. He found the statement, however, that all data were reported on the basis of ignited soil.

"That will reduce some of these amounts about one-tenth," he said to himself. "In our physics work in college, good soils generally lost about ten per cent. in weight by ignition, even after all hygroscopic moisture had been expelled; but these very poor soils haven't much to lose, I guess. They surely contain no carbonates and very little organic matter, although they may contain some combined water."

CHAPTER XXIV

THE NATION'S CAPITOL

PERCY spent three days in Washington.

"If I lived here long," he wrote his mother, "I think I should become as optimistic as the Secretary of Agriculture, even though the total produce of the original thirteen states should supply a still smaller fraction of the necessities of life required by their population. The Congressional Library is by far the finest structure I have ever seen. I cannot help feeling proud that I am an American when I walk through its halls and look upon the portraits of the great men who helped to make our country truly great.

"As I shook hands with the President of the United States at one of his public receptions held in the 'East Room' of the White House, I wondered if there was another country on the earth where the humblest subject could thus come face to face with the head of a mighty nation. In the Treasury Building I was permitted to join a small party of some distinction and shared with each of them the privilege of holding in my hands for a moment eight million dollars in government bonds.

"I have visited many of the great buildings, the Capitol, of course, and Washington's monument, which rises to a height of 555 feet above the surrounding land, or practically 600 feet above low-water level in the Potomac. There are many smaller monuments erected in honor of American heroes in various squares, circles, and parks throughout the City.

"The zoological garden took a full half-day, and I could have spent a much longer time there. They told me of a frightful occurrence that happened only last week. In a pool of water a very large alligator is kept confined by a low stout iron fence. A negro woman was leaning over the fence holding her baby in her arms and looking at the monster who seemed to be asleep; when, without a moment's warning, he thrust himself half out of the water and snapped the baby from her arms, swallowing it at one gulp as he settled back into the water. I fear the report is true enough, for they have made the fence higher in a very temporary manner, and I heard it mentioned by a dozen or more.

"I leave Washington by boat at five o'clock this afternoon, and I expect to land at Leonardtown, St. Mary county, Maryland, about six o'clock in the morning, when the boat will be ready to leave that port. It is a freight boat and stops for hours at large towns.

"I am planning for a trip into New England next week. I did not realize how easy it is to go there until I looked up the train service. In less than twelve hours' time, one can make the trip from the Virginia line, through the District of Columbia, Maryland, Delaware, Pennsylvania, New Jersey, New York, Connecticut, Rhode Island, and into Massachusetts,—ten different states, including the District. The trip from Galena to Cairo can hardly be made in so short a time, not even on the limited Illinois Central trains."

An hour before leaving the Washington hotel Percy chanced to meet a Congressman whom he had seen on several occasions at the University and who had spoken at the alumni banquet at the time of Percy's graduation.

"I'm very glad you introduced yourself, Mr. Johnston," said he. "Want to get a place down here, do you? Very likely I can help you some. I've helped several friends of mine to get good places. What are you after ?"

"I am thinking of getting a place of about three hundred acres," said Percy, "and I shall certainly appreciate any assistance or information you can give me."

"Whe-e-ew. What are you up to? Want to sell us a site for the new Government insane hospital, or going to lay out another addition to the city?"

"Neither," replied Percy. "I am looking for a piece of cheap land that I can build up and make into a good farm."

"Oh, ho!" said the Congressman. "That's it, is it? Well, now let me tell you that you've struck the wrong neck of the woods to find land that you can make a good farm out of. The land about here is cheap enough all right—cheaper than the votes of some politicians, but it can't be built up into good farms. Don't attempt the impossible, my friend. If you want cheap land for town sites or insane hospitals, right here's the country to land in; but if you want a good farm, you stay right in Illinois, or else follow Horace Greeley's advice and 'go West.'. That's a good suggestion for you, too. Just go West and get three hundred and twenty acres of the richest soil lying out of doors."

"There is not much land left in the West where the rainfall is sufficient for good crops," said Percy.

"Then take irrigated land. The Government is getting under way some big irrigation projects, and you ought to get in on the ground floor on one of those tracts. It is a fact that the apples from some of those irrigated farms sometimes bring more than $500 an acre."

"I don't doubt that," said Percy. "An illustration or example can usually be found to prove almost anything. I know that the Perrine Brothers, who conduct a fruit farm down in 'Egypt,' actually received $800 per acre for the apples grown on thirteen acres one year; and there is plenty of such land in Egypt that can be bought for less than $40 an acre, and near to the great markets. I am told, however, that there are from a dozen to a hundred applicants for every farm opened to settlement in the West in these years, and it is estimated that all of the arid lands that can ever be put under irrigation in the United States will provide homes for no more than our regular increase in population in five years, and that the only other remaining rich lands—the swamp areas—will be occupied by the increase of ten years in our population. It has seemed to me that it is high time we came back to these partially worn-out Eastern lands and begin to build them up. Here the rainfall is abundant, the climate is fine, and the markets are the best, and there are millions of acres of these Eastern lands that lie as nicely for farming as the Western prairies. Why should they not be built up into good farms?"

"Now, let me give you a little fatherly advice," said the Congressman, laying his hand on Percy's shoulder. "I tell you this land never was any good. If the East and South hadn't been settled first, they never would have been settled. Poor land remains poor land, and good land remains good land; and if you want to farm good land, you better stay right in the corn belt. You can't grow anything on these Eastern lands without fertilizer and the more you fertilize the more you must, and still the land remains as poor as ever. Just leave off the fertilizer one year and your crop is not worth harvesting. These lands never were any good and they never will be."

"But that is hardly in accord with what the people now living on these old Eastern farms report for the conditions of agriculture in the times of their ancestors."

"Oh, yes, I know people are always talking about their ancestors, and especially Virginians; but, Caesar! I wonder what their ancestors would think of them! You can't afford to take any stock in the ancestry of these old Virginians."

"I call to mind that the historical records give much information along this line," said Percy. "It is recorded that mills for grinding corn and wheat were common, that the flour of Mount Vernon was packed under the eye of Washington, and we are told that barrels of flour bearing his brand passed in the export markets without inspection. History records that the plantations of Virginia usually passed from father to son, according to the law of entail, and that the heads of families lived like lords, keeping their stables of blooded horses and rolling to church or town in their coach and six, with outriders on horseback. Their spacious mansions were sometimes built of imported brick; and, within, the grand staircases, the mantles, and the wainscot reaching from floor to ceiling, were of solid mahogany, elaborately carved and paneled. The sideboards shone with gold and silver plate, and the tables were loaded with the luxuries from both the New and the Old World, and plenty of these old mansions still exist in dilapidated condition."

"That all sounds good for history," said the Congressman, "but the historian probably got his information from some of these old Virginians whose only religion is ancestral worship. If the lands were ever any good they'd be good now. Good lands stay good. As an Illinois man, you ought to know that. My father settled in Illinois and I tell you his land is better to-day than it was the day he took it from the Government."

"My grandfather also took land from the Government," said Percy, "but the land that he first put under cultivation is not producing as good crops now as it used to, even though—"

"Then it must be you don't farm it right. Of course you don't want to corn your land to death. I lived on the farm long enough to learn that; but if you'll only grow two or three crops of corn and then change to a crop of oats, you'll find your land ready for corn again; and, if you'll sow clover with the oats and plow the clover under the next spring, you'll find the land will grow more corn than ever your grandfather grew on it."

"But how can we maintain the supply of plant food in the soil by merely substituting oats for corn once in three or four years and turning under perhaps a ton of clover as green manure. That amount of clover would contain no more nitrogen than 40 bushels of corn would remove from the soil, and of course the clover has no power to add any phosphorus or other mineral elements."

"Oh, yes. I've heard all about that sort of talk. You know I'm a U. of I. man myself. I studied chemistry in the University under a man who knew more in a minute than all the 'tommy rot' you've been filled up with. I also lived on an Illinois farm, and I speak from practical experience. I know what I am talking about, and I don't care a rap for all the theories that can be stacked up by your modern college professor, who wouldn't know a pumpkin if he met one rolling down hill. I tell you God Almighty never made the black corn belt land to be worn out, and he doesn't create people on this earth to let 'em starve to death. Don't you understand that?"

"I am afraid that I do not," replied Percy. "I have received no such direct communication; but I saw a letter written from China by a missionary describing the famine-stricken districts in which he was located. He wrote the letter in February and said that at that time the only practical thing to do in that district was to let four hundred thousand people starve and try to get seed grain for the remainder to plant the spring crops. I have a "Handbook of Indian Agriculture" written by a professor of agriculture and agricultural chemistry at one of the colleges in India. I got it from one of the Hindu students who attended the University when I was there. This book states that famine, local or general, has been the order of the day in India, and particularly within recent years. It also states that in one of the worst famines in India ten million people died of starvation within nine months. The average wage of the laboring man in India, according to the Governmental statistics, is fifty cents a month, and in famine years the price of wheat has risen to as high as $3.60 a bushel. This writer states that the most recent of all famines; namely, that prevailing in most parts of India from 1897 to 1900, was severer than the famine of 1874 to 1878. No, Sir, I am not sure that I understand just what God's intentions are concerning the corn belt, but it is recorded that the Lord helps him who helps himself, and that man should earn his bread by the sweat of his brow. If God made the common soil in America with a limited amount of phosphorus in it, He also stored great deposits of natural rock phosphate in the mines of several States, and perhaps intended that man should earn his bread by grinding that rock and applying it to the soil. Possibly the Almighty intended—"

"Now, I'm very sorry, Mr. Johnston, but I have an engagement which I must keep, and you'll have to excuse me just now. I'm mighty glad to have met you and I'd like to talk with you for an hour more along this line; but you take my advice and stick to the corn belt land. Above all, don't begin to use phosphates or any sort of commercial fertilizer; they'll ruin any land in a few years; that's my opinion. But then, every man has a right to his own opinion. and perhaps you have a different notion. Eh?"

"I think no man has a right to an opinion which is contrary to fact," Percy replied. "This whole question is one of facts and not of opinions. One fact is worth more than a wagonload of incorrect opinions. But I must not detain you longer. I am very glad to have met you here. In large measure the statesmen of America must bear the responsibility for the future condition of agriculture and the other great industries of the United States, all of which depend upon agriculture for their support and prosperity. Good bye."

"I'll agree with you there all right; the farmer feeds them all. Good bye."

CHAPTER XXV

A LESSON ON TOBACCO

PERCY found Leonardtown almost in the center of St. Mary county, situated on Breton bay, an arm of the lower Potomac.

From the data recorded on the back of his map of Maryland, Percy noted that a population of four hundred and fifty-four found support in this old county seat, according to the census of I 900. After spending the day in the country, he found himself wondering how even that number of people could be supported, and then remembered that there is one industry of some importance in the United States which exists independent of agriculture, an industry which preceded agriculture, and which evidently has also succeeded agriculture to a very considerable extent in some places; namely, fishing.

"Clams, oysters and fish, and in this order," he said to himself, "apparently constitute the means of support for some of these people."

And yet the country was not depopulated, although very much of the arable land was abandoned for agricultural purposes. A farm of a hundred acres might have ten acres under cultivation, this being as much as the farmer could "keep up," as was commonly stated. This meant that all of the farm manure and other refuse that could be secured from the entire farm or hauled from the village, together with what commercial fertilizer the farmer was able to buy, would not enable him to keep more than ten acres of land in a state of productiveness that justified its cultivation. Tobacco, corn, wheat and cowpeas were the principal crops. Corn was the principal article of food, with wheat bread more or less common. The cowpeas and corn fodder usually kept one or more cows through the winter when they could not secure a living in the brush. Tobacco, the principal "money crop," was depended on to buy clothing, and "groceries," which included more or less fish and pork, although some farmers "raised their own meat," in part by fattening hogs on the acorns that fell in the autumn from the scrub oak trees.

One farm of one hundred and ninety acres owned by an old lady, who lived in the nearby country village was rented for $100 a year, which amounted to about fifty-two and one-half cents an acres as the gross income to the landowner. After the taxes were paid, about thirty cents an acre remained for repairs on buildings and fences and interest on the investment.

Percy spent some time on a five hundred acre farm belonging to an old gentleman who still gave his name as F. Allerton Jones, a man whose father had been prominent in the community. According to the county soil map which had been presented to Percy by the Bureau of Soils, the soil of this farm was all Leonardtown loam, except about forty acres which occupied the sides of a narrow valley a bend of which cut the farm on the south side.

"My father had this whole farm under cultivation," said Mr. Jones, "except the hillsides. But what's the use? We get along with a good deal less work, and I've found it better to cultivate less ground during the forty odd years I've had to meet the bills. But I've kept up more of my land than most of my neighbors. I reckon I've got about eighty acres of good cleared land yet on this farm, and the leaves and pine needles we rake up where the trees grow on the old fields make a good fertilizer for the land we aim to cultivate, and I get a good many loads of manure from friends who live in the village and keep a cow or a horse.

"The last crop I raised on that east field, where you see those scrub pines, was in 1881. I finished cultivating corn there the day I heard about President Garfield being shot; and it was a mighty hot July day too. My neighbor, Seth Whitmore, who died about ten years ago, came along from the village and waited for me to come to the end of the row down by the road and he told me that Garfield was shot. We both allowed the corn would be a pretty fair crop and when I gathered the fodder that fall there was a right smart of a corn crop. Yes, Sir, it's pretty good land, but we don't need much corn, no how, and we can make more money out of tobacco. Of course it takes lots of manure and fertilizer to grow a good patch of tobacco, but good tobacco always brings good money."

"About how much money do you get for an acre of tobacco?" asked Percy.

"That varies a lot with the quality and price—sometimes $100—sometimes $300, when the trust don't hold the price down on us. We can raise good tobacco and good tobacco brings us good money. We can always manure an acre or two for tobacco and get our groceries and some clothes now and then, and that's about all anybody gets in this world, I reckon. But taxes are mighty high, I tell you. About $75 to $80 I have to pay. Are taxes high out West?"

"We pay about forty to fifty cents an acre in the corn belt," Percy replied; "but, in a course I took in economics, I learned that the taxes do not vary in proportion to land values. Poor lands, if inhabited, must always pay heavy taxes; whereas, large areas of good land carry lighter taxes compared with their earning capacity. You must provide your regular expenses for county officers, county courthouse, jail, and poorhouse, about the same as we do. Your roads and bridges cost as much as ours; and the schools in the South must cost more than ours, for a complete double system of schools is usually provided.

"But did you say that you paid fifty cents an acre in taxes?" asked Mr. Jones.

"Yes, about that, in the corn belt," replied Percy, "but not so much in Southern Illinois where the land is poor. I think the farmers in that section pay taxes as low as yours. Perhaps twenty cents an acre."

"Do you mean to say that you have poor land in Illinois?"

"Yes, the common prairie land of Southern Illinois must be called poor as compared with the corn belt land. There is a good deal of land in Southern Illinois that was put under cultivation before 1820, and eighty crops must have made a heavy draft upon the store of plant food originally contained in those soils."

"Only since 1820? Why, we began to till the soil right here, Young Man, in St. Mary County, in 1634 and don't you know, Sir, that we had a rebellion here as early as 1645? Yes, Sir, that was one hundred and seventy-five years before 1820. So you've raised only eighty crops and the land is already getting poor, and we've raised two hundred and fifty crops—well, maybe, not quite so many, for we've been giving our land a good deal of rest for the last fifty or sixty years; but my grandfather used to raise twenty-five bushels of wheat to the acre with the help of a hundred pounds of land-plaster, and I've no doubt I could do it again today if I cared to raise wheat, but one acre of tobacco is worth ten of wheat, so why should I bother with wheat?"

"Twenty-five bushels of wheat per acre," repeated Percy, half to himself. "The total supply of phosphorus still remaining in the plowed soil would be sufficient for only twenty more crops like that. Two hundred years of such crops would require 1600 pounds of phosphorus, making nearly 1800 pounds at the beginning, if it all came from the plowed soil. That is one and a half times as much as is now contained in our common corn belt prairie land."

"More stuff in our land than in yours, did you say?" questioned the old man. "I told you we had pretty good soil here, but I've always allowed your soil was better, but maybe not. I tell you manure lasts on this land. You can see where you put it for nigh twenty years. Then we rest our land some and that helps a sight, and if the price stays up we make good money on tobacco. I'm sorry your land is getting so poor out West, especially if you can't raise tobacco. Ever tried tobacco, Young Man?—gosh, but you remind me of one of them Government fellows who came driving along here once when Bob and his brothers were plowing corn right here about three years ago. Bob's my tenant's nigger, and he ain't no fool either, even if he is colored; but then, to tell the truth, he ain't much colored. Well, I was sitting under a tree right here smoking and keeping an eye on the niggers unbeknownst to them when one of them

Government fellows stopped his horse as Bob was turning the end, and says he to Bob:

"'Your corn seems to be looking mighty yellow?'

"'Yes, suh,' says Bob. 'Yes, suh, we done planted yellow corn.'

"'Well, I mean it looks as though you won't get more than half a crop,' says he.

"'I reckon not,' says Bob. 'The landlord, he done gets the other half.'

"With that the fellow says to Bob:

"'It seems to me you're mighty near a fool.'

"'Yes, suh,' says Bob, 'and I'm mighty feared I'll catch it if I don't get a goin'.'

"The fellow just gave his horse a cut and drove on, but I liked to died. He'd been here two or three times pestering me with questions about raising tobacco. Say, you ain't one of them Government fellows, are you? They were travelling all around over this county three years ago, learning how we raised tobacco and all kinds of crops. They had augers and said they were investigating soils, but I never heard nothing of 'em since. Have you got an auger to investigate soils with?"

Percy was compelled to admit that he had an auger and that he was trying to learn all he could about the soil.

He had driven to Mr. Jones' farm because his land happened to be situated in a large area of Leonardtown loam, and he felt free to stop and talk with him because he had found him leaning against the fence, smoking a cob pipe, apparently trying to decide what to do with some small shocks of corn scattered over a field of about fifteen acres.

Percy stepped to the buggy and drew out his soil auger, then returned to the corn field and begun to bore a hole near where Mr. Jones was standing.

"That's the thing," said he, "the same kind of an auger them fellows had three years ago. Still boring holes, are you? Want to bore around over my farm again, do you?"

Percy replied that he would be glad to make borings in several places in order that he might see about what the soil and subsoil were like in that kind of land.

"That's all right, Young Man. Just bore as many holes as you please. I suppose you'd rather do that than work; but you'll have to excuse me. I've got a lot to do today, and it's already getting late. I can't take time again to tell you fellows how to raise tobacco. Good day."

CHAPTER XXVI

ANOTHER LESSON ON TOBACCO

THE old man had stuck his cob pipe in a pocket and filled his mouth with a chew of tobacco.

He walked by Percy's buggy with the tobacco juice drizzling from the corners of his mouth, and turned down the road toward the house.

Percy finished boring the hole and then returned to the buggy.

"Christ, that old man eats tobacco like a beast!" exclaimed the driver as Percy approached.

"Are you speaking to me?" asked Percy.

"Why, certainly."

"That is not my name, please," admonished Percy, "but I can tell you that I know Him well and that He is my best friend."

"What, old Al Jones?"

"No,—Christ," replied Percy, with a grieved expression plainly discernible.

"Oh," said the driver.

They drove past the Jones residence and out into the field beyond. The house one might have thought deserted except for the well-beaten paths and the presence of chickens in the yard. It was a large structure with two and a half stories. The cornice and window trimmings revealed the beauty and wealth of former days. Rare shrubs still grew in the spacious front yard, and gnarled remnants of orchard trees were to be seen in the rear. A dozen other buildings, large and small, occupied the background, some with the roofs partly fallen, others evidently still in use.

"How old do you suppose these buildings are?" asked Percy of the driver.

"About a hundred years," he replied, "and I reckon they've had no paint nor fixin' since they was built, 'cept they have to give some of 'em new shingles now and then or they'd all fall to pieces like the old barns back yonder."

Percy examined the soil in several places on the Jones farm and on other farms in the neighborhood. They lunched on crackers and canned beans at a country store and made a more extended drive in the afternoon.

"It is a fine soil," Percy said to the driver, as they started for Leonardtown. "It contains enough sand for easy tillage and quick drainage, and enough clay to hold anything that might be applied to it."

"That's right," said the driver. "Where they put plenty of manure and fertilizer they raise tobacco three foot high and fifteen hundred pounds to the acre, but where they run the tobacco rows beyond the manured land so's to be sure and not lose any manure, why the stuff won't grow six inches high and it just turns yellow and seems to dry up, no matter if it rains every day. Say, Mister, would you mind telling me if you're a preacher?"

"Oh, no," replied Percy, " — I am not a preacher, any more than every Christian must be loyal to the name."

"Well, anyway, I've learned a lesson I'll try to remember. I never thought before about how it might hurt other people when I swear. I don't mean nothing by it. It's just a habit; but your saying Christ is your friend makes me feel that I have no business talking about anybody's friend, any more than I'd like to hear anybody else use my mother's name as a by-word. I reckon nobody has any right to use Christ's name 'cept Christians or them as wants to be Christians.

I reckon we'd never heard the name if it hadn't a been for the Christians.

"But I don't have so many bad habits. I don't drink, nor smoke, nor chew; and I don't want to. My father smoked some and chewed a lot, and I know the smell of tobacco used to make my mother about as sick as she could be; but she had to stand it, or at least she did stand it till father died; and now she lives with me, and I'm mighty glad she don't have to smell no more tobacco

"She often speaks of it—mother does; and she says she's so thankful she's got a boy that don't use tobacco. She says men that use tobacco don't know how bad it is for other folks to smell 'em. Why, sometimes I come home when I've just been driving a man some place in the country, riding along like you and I are now, and he a smoking or chewing, or at least his clothes soaked full of the vile odor; and when I get home mother says, 'My! but you must have had an old stink pot along with you to-day.' She can smell it on my clothes, and I just hang my coat out in the shed till the scent gets off from it.

"No, Sir, I don't want any tobacco for me, and I don't know as I'd care to raise the stuff for other folks to saturate themselves with either; and every kid is allowed to use it nowadays, or at least most of them get it. It's easy enough to get it. Why, a kid can't keep away from getting these cigarettes, if he tries. They're everywhere. Every kid has hip pockets full; and I know blamed well that some smoke so many cigarettes they get so they aren't more than half bright. It's a fact, Sir,—plenty of 'em too; and some old men, like Al Jones, are just so soaked in tobacco they seem about half dead. Course it ain't like whiskey, but I think it's worse than beer if beer didn't make one want whiskey later.

"But as I was saying, I feel that I have no business saying things about,—about anybody you call your friend, and I think I'll just swear off swearing, if I can."

"You can if you will just let Him be your friend."

"Well, I don't know much about that," was the slow reply. "That takes faith, and I don't have much faith in some of the church members I know."

"That used to trouble me also," said Percy, "until one time the thought impressed itself upon me that even Christ himself did all His great work with one of the twelve a traitor; and this thought always comes to me now when self-respecting men object to uniting with organized Christianity because of those who may be regarded as traitors or hypocrites, but not of such flagrant character as to insure expulsion from the Church?"

"Do you believe in miracles?" asked the driver.

"Oh, yes," said Percy, "in such miracles as the growth of the corn plant."

"Why, that isn't any miracle. Everybody understands all about that."

"Not everybody," replied Percy. "There is only One who understands it. There is only one great miracle, and that is the miracle of life. It is said that men adulterate coffee, even to the extent of making the bean or berry so nearly like the natural that it requires an expert to detect the fraud; but do you think an imitation seed would grow?"

"No, it wouldn't grow," said the driver.

"Not only that," said Percy, "but we may have a natural and perfect grain of corn and it can never be made to grow by any or all of the knowledge and skill of men, if for a single instant the life principle has left the kernel, which may easily result by changing its temperature a few degrees above or below the usual range. The spark of life returns to God who gave it, and man is as helpless to restore it as when he first walked the earth.

"What miracle do you find hard to accept?" asked Percy.

"How could Jesus know that Lazarus had died when he was on the other side of the mountain?"

"I don't know," Percy replied; "perhaps by some sort of wireless message which his soul could receive. I don't know how, but it was no greater miracle than it would have been then to have done what I did last week."

The driver turned to look squarely at Percy as though in doubt of his sanity, but a kindly smile reassured him.

"Our train coming into Cincinnati ran in two sections," Percy continued, "and the section behind us was wrecked, three travellers being killed and about fifteen others wounded. I was sure my mother would hear of the wreck before I could reach her with a letter, and so I talked with her from Cincinnati over the long distance 'phone, with which we have always had connection since I first went away to college. Yes, I talked with her, and, though separated by a distance three times the entire length of Palestine, I distinctly heard and recognized my mother's voice. Oh, yes, I believe in miracles; but that is a matter of small consequence. The important thing is that we have faith in God and faith in Jesus Christ, his Son."

"Well, that's what troubles me," said the driver. "How's one to get faith?"

"There are two methods of receiving faith," replied Percy. "Faith cometh by prayer." "Yes, Sir, I believe that." "And, faith cometh by hearing." "Hearing what?" "Hearing by the Word of God; hearing those who have studied His Word and who testify of Him; and hearing with an ear ready to receive the truth."

CHAPTER XXVII

EIGHTEEN TO ONE

TWO days later Percy was in Rhode Island visiting a farm owned by Samuel Robbins, one of the most progressive and successful farmers of that State.

Mr. Robbins' farm lay in what appeared to be an ancient valley, several miles in width, although only a small stream now winds through it to the sea seven miles away.

"So you are from Illinois," said Mr. Robbins, after Percy had introduced himself and explained the nature of his visit. "The papers have a good deal to say about the corn you grow in Illinois; but have you noticed that the Government reports show our average yield of corn in New England is higher than yours in Illinois?"

"Yes, Sir," Percy replied, "I have noticed that and I have come to Rhode Island to learn how to raise more corn per acre. I have noticed, however, that New England corn does not occupy a large acreage."

Well, now, we count corn as one of our big crops, next to hay. You'll see plenty of corn fields right here in Rhode Island."

"Would you believe that we actually raise more corn on one farm in Illinois than the total corn crop of Rhode Island?"

"You don't tell!"

"Yes," said Percy, "the Isaac Funk farm in McLean County grows more corn on seven thousand acres a year, with an average yield certainly above fifty bushels per acre, and surely making a total above 350,000 bushels; while the State of Rhode Island grows corn on nearly ten thousand acres with an average yield of thirty-two bushels, making a total yield of about 320,000 bushels."

"Well, I'll give it up; but I'd like to know how much corn you raise in the whole State of Illinois."

"Our average production," said Percy, "is about equal to the total production of Maine, New Hampshire, Vermont, Massachusetts, Rhode Island, Connecticut, New York, New Jersey, Pennsylvania, Delaware, Maryland, Virginia, North and South Carolina, Georgia, Florida, Alabama, and Mississippi."

"Eighteen of us!" exclaimed Mr. Robbins, who had counted on his fingers from New York to Mississippi. "And you come to Rhode Island to learn how to raise corn?"

"Yes, I came to learn how you raise more than thirty-five bushels of corn per acre as an average for New England, while we raise less than thirty-five bushels as an average in Illinois, and while Georgia, a larger State than Illinois, raises only eleven bushels per acre as a ten year average. Illinois is a new State, but I call to mind that Roger Williams settled in Rhode Island in 1636 and that he was joined by many others coming not only from Massachusetts but also from other sections. I assume that much of the land in Rhode Island has been farmed for 250 years, and the fact that you are still producing more than thirty bushels of corn per acre, as an average, is, it seems to me, a fact of great significance. I suppose you use all the manure you can make from the crops you raise and perhaps use some commercial fertilizer also. I should like to know what yield of corn you produce without any manure or fertilizer?"

"We don't produce any," said Mr. Robbins; "at least we know we wouldn't produce any corn without fertilizing the land in one way or another. If you will walk over here a little ways you can see for yourself. I didn't have quite enough manure to finish this field and I had no more time to haul seaweed so I planted without getting any manure on a few rods in one corner, and the corn there wouldn't make three bushels from an acre. I didn't bother to try to cut it, but the cows will get what little fodder there is as soon as I can get the shocks out of the field and turn 'em in for a few days to pick up what they can."

Percy examined the corn plants still standing in the corner of the field. They had grown to a height of about two feet. Most of them had tassels and many of them appeared to have little ears, but really had only husks containing no ear. In a few places where the hill contained only one plant a little nubbin of corn could be found.

"I don't mean to let any of my land get as poor as this field was," continued Mr. Robbins, "but I just couldn't get to it, and I left it in hay about two years longer than I should have done. Last year was first class for hay but this field had been down so long it was hardly worth cutting."

"About what yield do you get from the manured land?" inquired Percy.

"In a fair year I get about forty bushels, and that's about what I am getting this year from my best fields. You see there's lots of corn in these shocks. There's about an average ear, and we get five or six ears to the hill."

"Eight-row flint," said Percy, as he took the ear in his hand and drew a celluloid paper knife from his vest pocket with a six-inch scale marked on one side.

"Yes, Sir, our regular Rhode Island White Cap."

"Just five inches long. Weight about three ounces?"

"Perhaps. We count on about four hundred ears to the bushel. If we get four thousand hills to the acre one ear to the hill would give us ten bushels per acre, so you see we only have to have four ears to the hill to make our forty bushels. A good many hills have five to six ears, but then of course, some hills don't have much of any, so I suppose my corn makes an average of four ears about like that."

"I suppose you feed all of the corn you raise in order to produce as much manure as possible."

"Feed that corn! Not much we don't. Why, corn like that brings us close on to a dollar a bushel. No, Sir, we don't feed this corn. It's all used for meal. It makes the best kind of corn meal. No, we buy corn for feed; western corn. Oh, we feed lots of corn; three times as much as we raise; but we don't feed dollar corn, when we can buy western corn for seventy-five or eighty cents.

"I sell corn and I sell potatoes; that's all except the milk. I keep most of my land in meadow and pasture and feed everything I raise except the corn and potatoes. And milk is a good product with us. We average about sixty cents a pound for butter fat, and it's ready money every month; and, of course, we need it every month to pay for feed."

"Then you produce on the farm all the manure you use," suggested Percy, "but I think you mentioned hauling seaweed."

"Yes, and I haul some manure, too, when I can get it; but usually there are three or four farmers ready to take every load of town manure."

"You get it from town for the hauling?"

"Well, I guess not," said Mr. Robbins emphatically and with apparent astonishment at such a question. "I don't think I would haul seaweed seven miles if I could get manure in town for nothing. Manure is worth $1.50 a ton Iying in the livery stable, and there are plenty to take it at that right along. I'd a little rather pay that than haul seaweed; but the manure won't begin to go around, and so there's nothing left for us but seaweed; and, if we couldn't get that, the Lord only knows what we could do."

"How much seaweed can you haul to a load, and about how many loads do you apply to the acre?"

"When the roads are good we haul a cord and a quarter, and we put ten or twelve loads to the acre for corn and then use some commercial fertilizer."

"Do you know how much a cord of the seaweed would weigh?"

"Yes, a cord weighs about a ton and a half."

"Then you apply about twenty tons of seaweed to the acre for corn?"

"Yes, but some use less and some more; probably that's about an average. Hauling seaweed's a big job and a bad job. We have to start from home long before daylight so as to get there and get the weed while the tide is out, and then we get back with our load about two o'clock in the afternoon; and, by the time we eat and feed the team, and get the load to the field and spread, there isn't much time left that day, especially when you've got to pile out of bed about two o'clock the next morning and hike off for another load."

"Then you use some fertilizer in addition to the seaweed? May I ask how much fertilizer you apply to the acre and about how much it costs per ton?"

"Where we spread seaweed for corn, we add about four hundred and fifty pounds per acre of fertilizer that costs me $26 a ton, but I have the agency and get it some cheaper than most have to pay. Then for potatoes we apply about 1500 pounds of a special potato fertilizer that costs me $34 a ton."

"The fertilizer costs you about $6 an acre for the corn crop and $25 for potatoes," said Percy; "and then you have the cost of the seaweed. I should think you would need to count about $25 or $30 an acre for the expense of hauling seaweed."

"Yes, all of that if we had to pay for the work, but of course we can haul seaweed more or less when the farm work isn't crowding, and we don't count so much on the expense. It doesn't take the cash, except may be a little for a boy to drive one team when we haul two loads at a time; and we don't use seaweed for potatoes. The corn crop will generally more'n pay for it and the fertilizer too; and the seaweed helps for three or four years, especially for grass. There's good profit in potatoes, too, when we get a crop, but they're risky, considering the money we have to pay for fertilizer."

CHAPTER XXVIII

FARMER OR PROFESSOR

AFTER leaving Rhode Island, Percy spent two days in and about Boston, and then returned to Connecticut for a day. The weather had turned cold; the ground had frozen and the falling snow reminded him that it was the day before Thanksgiving.

From New London he took a night boat to New York, and then took passage on a Coast Line vessel from New York to Norfolk.

The weather had cleared and the wind decreased until it was scarcely greater than the speed of the ship.

Whether or not the dining room service was extraordinary because of the day, Percy was soon convinced that the only way to travel was by boat. He regretted only that his mother was not with him to enjoy that day. For hours they coasted southward within easy view of the New Jersey shore, dotted here and there with cities, towns, and villages. Light houses marked the rocky points where danger once lurked for the men of the sea.

The sea itself was of constant interest; and hundreds of craft were passed or met. Here a full-rigged sailing vessel lazily drifting with the wind; there a giant little tug puffing in the opposite direction with a string of barges in tow loaded almost to the water's edge.

Norfolk was reached early the next morning, and before noon Percy passed through Petersburg on his way to Montplain. He changed cars at Lynchburg and arrived at Montplain before dark. In accordance with a promise to Mr. West he had notified him of his plans.

Would Adelaide met him, and if so would she have the family carriage and again insist upon his riding in the rear seat? He had found these questions in his mind repeatedly since he left New London, with no very definite purpose before him except to arrive at Montplain at the appointed time.

Yes, it was the family carriage. He saw the farm team tied across the street from the depot. As he left the train he caught a glimpse of Adelaide standing with the group of people who were waiting to board the train. She extended her hand as he reached her side.

"Mr. Johnston, meet my cousin, Professor Barstow."

"I am glad to meet you, Professor," said Percy, as he shook hands with a tall young man about his own age. Percy noted his handsome face and gentlemanly bearing.

"Miss Adelaide calls me cousin," said Barstow, "because my aunt married her uncle."

"Well, Sir, if we're not cousins, then I'm Miss West and not Miss Adelaide. Is that too much for an absent-minded professor to remember?"

"I am afraid it is," said Barstow, "and I am sure I would rather be cousins."

"Professor Barstow leaves on this train," Adelaide explained to Percy; "excuse me, please."

Percy raised his hat as he stepped back from the crowd and waited for the parting of the two. He was sure that Barstow held her hand longer than was necessary, and he also noticed that her face flushed as she rejoined him after the train started.

"Will you take the rear seat?" she asked. as they reached the carriage.

"If you so prefer."

"That seat is for our guests, so I don't prefer," came her reply, which left Percy wholly in the dark as to her wishes.

"Then let me be your coachman rather than your guest."

"If you so prefer," she repeated, and without waiting for assistance quickly mounted to the front seat, leaving him to occupy the driver's seat beside her.

"Captain and Mrs. Stone of Montplain were with us for Thanksgiving and I came with the carriage to take them home. Professor Barstow has also been spending his Thanksgiving vacation visiting with papa."

"Thank you," said Percy, as he took the lines and turned the horses toward Westover.

"You are certainly welcome to drive this team if you enjoy it."

"I thank you for that also," said Percy. Adelaide noted the word _also, _but she only remarked that she hoped he had enjoyed his travels, though she could not understand what pleasure he could find in visiting old worn-out farms.

"Of all things," she continued, "it seems to me that farming is the last that anyone would want to undertake."

"It is both the first and the last," said Percy. "As you know, when our ancestors came to America, agriculture was the first great industry they were able to develop. Other industries and professions follow agriculture and must be supported in large measure by the agricultural industry. Merchants, lawyers, doctors and teachers are in a sense agricultural parasites."

An hour before he would not have included teachers in this class; for, next to the mother in the home, he felt that the teacher in the school is the greatest necessity for the highest development of the agricultural classes.

"Without agriculture," he continued, "America could never have been developed, and, unless the prosperity of American agriculture can be maintained, poverty is the only future for this great nation. The soil is the greatest source of wealth, and it is the most permanent form of wealth. The Secretary of Agriculture at Washington told me a few days ago that eighty-six per cent. of the raw materials used in all our manufacturing industry are produced from the soil.

"Yes, agriculture is certainly the first industry in this country; and I am fully convinced that to restore the fertility of the depleted soils of the East and South, and even to maintain the productive power of the great agricultural regions of the West, deserves and will require the best thought of the most influential people of America.

"Throughout the length and breadth of this land, the almost universal purpose of the farmers is to work the land for all they can get with practically no thought of permanency. The most common remark of the corn belt farmer is that his land doesn't show much wear yet; and it is holding up pretty well, or as well as could be expected; or that he thinks it will last as long as he does. All recognize that the land cannot hold up under the systems of farming that are being practiced, and these systems are essentially the same as have been followed in America since 1607. What the Southern farmer did with slave labor, the Western farmer is now doing with the gang plow, the two-row cultivator, and the four-horse disks and harrows. In addition he tile-drains his land which helps to insure larger crops and more rapid soil depletion. He even uses clover as a soil stimulant, and spreads the farm fertilizer as thinly as possible with a machine made for the purpose in order to secure both its plant food value and its stimulating effect. Positive soil enrichment is practically unknown in the great corn belt.

"Robbery is a harsh word; and yet the farmers and landowners of America are and always have been soil robbers; and they not only rob the nation of the possibility of permanent prosperity, but they even rob themselves of the very comforts of life in their old age and their children and grandchildren of a rightful inheritance.

"Worse than all this, or at least more lamentable, is the fact that it need not be. The soils of Virginia need not have become worn out and abandoned; because the earth and the air are filled with the elements of plant food that are essential to the restoration and permanent maintenance of the high productive capacity of these soils. Moreover there is more profit and greater prosperity for the present landowner in a possible practicable system of positive soil improvement than under any system which leads to ultimate depletion and abandonment of the land.

"The profit in farming lies first of all in securing large crop yields. It costs forty bushels of corn per acre in Illinois to raise the crop and pay the rent for the land or interest and taxes on the investment. With land worth $150 an acre, it will require $8 to pay the interest and taxes. Another $8 will be required to raise the crop and harvest and market it, even with very inadequate provision made for maintaining the productive power of the soil, such as a catch crop of clover, or a very light dressing of farm fertilizer. A forty-bushel crop of corn at forty cents a bushel, which is about the ten year average price for Illinois, would bring only $16 an acre, and this would leave no profit whatever.

"A crop of fifty bushels would leave only ten bushels as profit; but, if we could double the yield and thus produce a hundred bushels per acre, the profit would not be doubled only, but it would be six times as great as from the fifty bushel crop. In other words, 100 bushels of corn from one acre would yield practically the same profit as fifty bushels per acre from six acres, simply because it requires the first forty bushels from each acre to pay for the fixed charges or regular expense.

"It is not the amount of crop the farmer handles, but the amount of actual profit that determines his prosperity. It requires profit to build the new home or repair the old one, to provide the home with the comforts and conveniences that are now to be had in the country as well as in the city; to send the boys and girls to college; to provide for the expense of travel and the luxuries of the home."

Percy stopped himself with an apology.

"I hope you will pardon me, Miss West. I forget that this subject may be of no interest to you, and I have completely monopolized the conversation."

"I am glad you have told me so much," she replied. "I am deeply interested in what you have been saying. I never realized that agriculture could involve such very important questions in regard to our national prosperity. I only know that our farm has furnished us with a living but there has been very little of what you call profit. We children could never have gone away to school except that we were enabled to take advantage of some unusual opportunities. My brother almost earned his expenses as commissary in a boarding club at college. He felt that he could not come home for Thanksgiving because he had a chance to earn something and I have missed him so much. Most farmers get barely enough from their farms in these parts to furnish them a modest living and pay their taxes."

"That reminds me of your statement that farming is the last thing that you would expect anyone to undertake. In a large sense that is in accordance with the history of all great agricultural countries. After the great wave of easy spoilation of the land has passed, and the farmers reach a condition under which they need most of what they produce for their own consumption, the parasites are themselves forced to produce their own food. The lands become divided into smaller holdings and the agricultural inhabitants increase rapidly in proportion to the urban population which must depend upon the profits from secondary pursuits for a living. Thus ninety-five per cent. of the three hundred million people of India belong principally to the agricultural classes, and the farms of India average about two to three acres in size. Farming there is in no sense a profit-yielding business, but it is only a means of existence. The people live upon what they raise, so far as they can, although, as you must know, India is almost never free from famine. In Russia, the situation is but little better, for famine follows if the yield of wheat falls two bushels below the average. Special agents of the Bureau of Statistics of the United States Department of Agriculture report that at least one famine year occurs in each five year period, and sometimes even two; that the famine years are so frequent they are recognized as a permanent feature of Russian agriculture."

"But couldn't those poor starving people do some other kind of work and thus earn a better living?" asked Adelaide.

"No. Agriculture is the only hope," said Percy. "The soil is the breast of Mother Earth, from which her children must always draw their nourishment, or perish. It is the 'last thing,' as you truly said. Aside from hunting and fishing, there is no source of food except the soil, and, when this is insufficient for the people who produce it in the country, God pity the poor people who live in the cities. But let us not talk of this more. I ought not to have taken up the time of our ride through this beautiful scenery with a subject which tends always toward the serious. The leaves are all gone in New England, but here they have only taken on their most beautiful colors. 'What is so rare as a day in June?' could now well be answered, 'a day in November in Piedmont, Virginia.'"

"Do you know if your father received a letter for me from the chemist to whom I sent the soil samples?"

"Yes, it came in Wednesday's mail, and there is a letter from the University of Illinois and two others that Grandma says must be from a lady. Papa says he is anxious to know what results would be found in the chemist's report. May I listen while you tell papa about it? Indeed, I am extremely interested to know if anything can be done to make our farm produce such crops as it used to when grandmother was a little girl."

"Still I fear you will find it a very tiresome subject," said Percy. "It is, as a rule, not an easy matter to adopt a system of permanent improvement on land that has been depleted by a century or more of exhaustive husbandry. but you will be very welcome not only to listen but to counsel also. My mother can measure difficulties in advance better than most men; and I believe it is true that women will deliberately plan and follow a course involving greater hardship and privation than men would undertake. I cannot conceive of any man doing what my mother has done for me."

Adelaide glanced at Percy as he spoke of his mother. Something in his words or voice seemed to reveal to her a depth of feeling, a wealth of affection akin to reverence, such as she had never recognized before.

CHAPTER XXIX

THE ULTIMATE COMPARISON

WILKES was at the side gate to meet Adelaide and Percy, and the grandmother stood at the door as they reached the veranda.

"Lucky for us you got back before the Thanksgiving scraps are all gone," she said to Percy, "but I suppose even our Thanksgiving fare will be poor picking after you've been living in Washington and Boston."

"Even the Thanksgiving dinner on the boat was not equal to this," said Percy, as they sat down to the table loaded with such an a-bundance of good things as is rarely seen except on the farmer's table. The "scraps," if such there were, had no appearance of being left-overs, and there was monster turkey, browned to perfection and sizzling hot, placed before Mr. West ready for the carving knife.

Percy had opened the letter from the chemist, but said to Mr. West that it would take him an hour or more to compute the results to the form of the actual elements and reduce them to pounds per acre in order to make possible a direct comparison between the requirements of crops, on the one hand, and the invoice of the soil and application of plant food in manure and fertilizers, on the other hand.

"Please let me help you make the computations," said Adelaide, much to the surprise of her parents, who knew that she took no interest in affairs pertaining to farming. "I like mathematics and will promise not to make any mistakes if you will tell me how to do some of the figuring."

"Thank you," said Percy. "With your help it will take only half the time that I should require alone."

This proved to be correct, for in half an hour after supper they had the results in simplified form. Even the mother and grandmother joined the circle as Percy began to discuss the results with Mr. West

"Now here is the invoice," said Percy, "of the surface soil from an acre of land where we collected the first composite sample,—the land which you said had not been cropped since you could remember. This soil contains plant food as follows:

1,440 pounds of nitrogen 380 pounds of phosphorus 15,760 pounds of potassium 3,340 pounds of magnesium 10,420 pounds of calcium

"I'd like to know how these amounts compare with what your Illinois soil contains," said Mr. West.

"We have several different kinds of soil in Illinois," replied Percy. "The common corn belt prairie soil is called brown silt loam. It contains, as an average, 5000 pounds of nitrogen and 1200 pounds of phosphorus, or nearly four times as much of each of those elements as this Virginia soil which you say is too poor to cultivate.

"I wrote to the Illinois Experiment Station before I left Washington to see if I could get the average composition of the heavier prairie soil, which occupies the very flat areas that were originally swampy, and one of the letters you had received for me gives 8000 pounds of nitrogen and 2000 pounds of phosphorus as the general average for that soil. That is our most productive land, and it contains about five times as much of these two very important elements as your poorest land.

"Our more common Illinois prairie contains about 35,000 pounds of potassium, 9,000 pounds of magnesium, and I 1,000 pounds of calcium. This is more than twice as much potassium and nearly three times as much magnesium as in your poorest land, but the calcium content is about the same in your soil as in ours. However, as you will remember, your soil is distinctly acid and consequently markedly in need of lime, the magnesium and calcium evidently being contained in part in the form of acid silicates with no carbonates; whereas, our brown silt loam is a neutral soil and our black clay loam contains much calcium carbonate, the same compound as pure limestone."

"I am anxious to know about our best land," said Mr. West. "What did the chemist find in the soil from the slope where we get the best corn after breaking up the old pastures?"

"He found the following amounts in the surface soil," said Percy.

800 pounds of nitrogen

1,660 pounds of phosphorus

34, 100 pounds of potassium

8,500 pounds of magnesium

13,100 pounds of calcium

"Rich in everything but nitrogen," Percy continued, "richer than our common prairies in phosphorus and calcium, and nearly as rich in potassium and magnesium; but very, very poor in nitrogen. Legume plants ought to grow well on that land, because the minerals are present in abundance, and, while lack of nitrogen in the soil will limit the yield of all grains and grasses, there is no nitrogen limit for the legume plants if infected with the proper nitrogen-fixing bacteria, provided, of course, that the soil is not acid. You will remember, however, that even this sloping land is more or less acid, although here and there we found pieces of undecomposed limestone. With a liberal use of ground limestone, any legumes suited to this soil and climate ought to grow luxuriantly on those slopes."

"That reminds me that we are greatly troubled with Japan clover on those slopes," said Mr. West. "Of course it makes good pasture for a few months, but it doesn't come so early in the spring as blue grass and it is killed with the first heavy frost in the fall. We like blue grass much better for that reason, but when we seed down for meadow and pasture, the Japan clover always crowds out the timothy and blue grass on those slopes."

"And when you plow under the Japan clover, you get one or two good crops of grain," said Percy, "because this clover has stored up some much needed nitrogen and the soil is rich in all other necessary elements. Have you ever tried alfalfa on that kind of land? That is a crop that ought to do well there, especially if limestone were applied."

"Yes, I have tried alfalfa," replied Mr. West, "and I tried it on a strip that ran across one of those steep slopes; but it failed completely, and, as I remember it, it was poorer on that hillside than on the more level land."

"Did you inoculate it?" Percy asked.

"Inoculate it? No. I didn't do anything to it, but just sow it the same as I sow red clover."

"What does it mean to inoculate it?" asked Adelaide.

"It means to put some bugs on it," said the grandmother; "some germs or microbes, or whatever they are called. Don't you remember, Adelaide, that I told you about that when I read it in the magazine a while ago? Don't you remember that somebody was making it and a man could carry enough in his vest pocket to fertilize an acre and he wanted $2 a package. Charles said that $1.50 a hundred was more than he could afford to pay for fertilizer, and he didn't care to pay $2 for a vest pocket package. Isn't that the stuff, Mr. Johnston?"

"It listens like it, as the Swedes say," said Percy, "but the advertisements of these germ cultures put out by commercial interests are usually very misleading. The safest and best and least expensive method of inoculating a field for alfalfa is to use infested soil taken from some old alfalfa field or from a patch of ground where the common sweet clover, or mellilotus, has been growing for several years. I saw the sweet clover growing along the railroad near Montplain, and there is one patch on the roadside right where—when you enter the valley on the way to the station."

"Right where Adelaide smashed that nigger's eye with her heel and helped Mr. Johnston capture them both," broke in the grandmother. "That's the only good thing I can say for her peg heeled shoes."

Adelaide colored and Percy now understood what had been a puzzle to him.

"The same bacteria," he went on quickly, "live upon both the sweet clover and the alfalfa, or at least they are interchangeable. These bacteria are not a fertilizer in any ordinary sense, but they are more in the nature of a disease, a kind of tuberculosis, as it were; except that they do much more good than harm. They attack the very tender young roots of the alfalfa and feed upon the nutritious sap, taking from it the phosphorus and other minerals and also the sugar or other carbohydrates needed for their own nourishment, since they have no power to secure carbon and oxygen from the air, as is done by all plants with green leaves. On the other hand, these bacteria have power to take the free nitrogen of the air, which enters the pores of the soil to some extent, and cause it to combine with food materials which are secured from the alfalfa sap, and thus the bacteria secure for themselves both nitrogen and the other essential plant foods. The alfalfa root or rootlet becomes enlarged at the point attacked by the bacteria, and a sort of wart or tubercle is formed which resembles a tiny potato, as large as clover seed on clover or alfalfa, and, singularly, about as large as peas on cowpeas or soy beans. On plants that are sparsely infected, these tubercles develop to a large size and often in clusters. While the bacteria themselves are extremely small and can be seen only by the aid of a powerful microscope, the tubercles in which they live are easily seen, and they are sufficient to enable us to know whether the plants are infected."

"I wish you would tell me the difference between the words inoculated and infected," said Adelaide.

"Inoculated is used in the active sense and infected in the passive," said Percy. "Thus the red clover growing in the field is infected if there are tubercles on its roots, although it may never have been inoculated; and we inoculate alfalfa because it would not be likely to become infected without direct inoculation."

"Under favorable conditions," continued Percy, "these bacteria multiply with tremendous rapidity, somewhat as the germs of small pox or yellow fever multiply if allowed to do so. A single tubercle may contain a million germs which if distributed uniformly over an acre would furnish more than twenty bacteria for every square foot."

"There, Charles," said the grandmother, "wouldn't a vest pocketful of those bugs or germs be a big enough dose for one acre?"

"Well, but they're not a fertilizer, Mother," said Mr. West, "and besides Mr. Johnston says it is better to use the infected sweet clover soil and there is no need of paying $2 an acre for something we knew nothing about, and especially on land that is not worth more than $2 an acre."

"I don't care what it's worth," she replied, "some of it cost your grandfather $68 an acre, and it will never be sold for any $2, while I have any say so about it."

They waited for Percy to proceed.

"The individual bacteria are very short-lived," he continued, "and products of decay soon begin to accumulate in the tubercles. These products contain, in combined form, nitrogen which the bacteria have taken from the air, and in this form it is taken from the tubercles and absorbed through the roots into the host plant and thus serves as a source of nitrogen for all of the agricultural legumes.

"It should be kept in mind, of course, that the red clover has one kind of nitrogen-fixing bacteria, that the cowpea has a different kind, and that the soy bean bacteria are still different, while a fourth kind lives on the roots of alfalfa and sweet clover."

"How much infected sweet clover soil would I need to inoculate an acre of land for alfalfa?" asked Mr. West.

"If the soil is thoroughly infected, a hundred pounds to the acre will do very well if applied at the same time the alfalfa seed is sown and immediately harrowed in with the seed. If allowed to lie for several hours or days exposed to the sunshine after being spread over the land the bacteria will be destroyed, for like most bacteria, such as those which lurk in milk pails to sour the milk, they are killed by the sunshine."

" That's right," said the grandmother. "That's the way to sterilize milk pails and pans and crocks. I like crocks better than pans. They don't have any sort of joints to dig out."

"Of course," continued Percy, "a wagon load of infected soil will make a more perfect inoculation than a hundred pounds, and where it costs nothing but the hauling it is well to use a liberal amount."

"How deep should it be taken?" asked Mr. West.

"About the same depth as you would plow. The tubercles are mostly within six or eight inches of the surface. The bacteria depend upon the nitrogen of the air and this must enter the surface soil. Sometimes in wet weather the tubercles can be found almost at the surface of the ground, and when the ground cracks one can often find tubercles sticking out in the cracks an inch or two beneath the surface but protected from direct sunshine.

"These bacteria have power to furnish very large amounts of nitrogen to such a crop as alfalfa. The Illinois Station reports having grown eight and one-half tons of alfalfa per acre in one season. It was harvested in four cuttings. The hay itself was worth at least $6 a ton above all expenses, which would bring $51 an acre net profit for one year. Of course this was above the average, which is only about four and one-half tons over a series of several years. But suppose you can save only three tons and get $6 a ton net for it, as you could easily do by feeding it to your cattle and sheep. That would bring $18 an acre or six per cent. interest on $300 land. I am altogether confident that this could be done on your sloping hillsides, with their rich supplies of phosphorus and other mineral foods, provided, of course, that you use plenty of ground limestone and thoroughly inoculate the soil."

"Well, I shall certainly try alfalfa again," said Mr. West, "and if I can grow such crops of alfalfa as you think on the hillsides, I can have much more farm manure produced for the improvement of the rest of the land. By the way what did that chemist find in that sample you took of the other land where it does not wash so much as on the steeper slopes."

"He found the following:

1,030 pounds of nitrogen 1,270 pounds of phosphorus 16,500 pounds of potassium 7,460 pounds of magnesium 16,100 pounds of calcium

"Well, the phosphorus is not so low," said Mr. West.

"Fully equal to that in our $150 Illinois prairie," replied Percy, "and again the calcium is more than ours, with magnesium not far below, and potassium half our supply. Nitrogen is plainly the most serious problem on most of this farm, and limestone and legumes must solve that problem if properly used."

"Do you think this land could be made as valuable as the Illinois land just by a liberal use of limestone and legumes?" asked Adelaide.

"I should have some doubt about that," Percy replied. "Your very level uplands that neither lose nor receive material from surface washing are very deficient in phosphorus and much poorer than ours in potassium and magnesium; and your undulating and steeply sloping lands are more or less broken, with many rock outcrops on the points and some impassable gullies, which as a rule compel the cultivation of the land in small irregular fields. A three-cornered field of from two to fifteen acres can never have quite the same value per acre as the land where forty or eighty acres of corn can be grown in a body with no necessity of omitting a single hill. Then there is some unavoidable loss from surface washing, so that to maintain the supply of organic matter and nitrogen will require a larger use of legumes than on level land of equal richness. In addition to this is the initial difference in humus content. This is well measured by the nitrogen content. While your soil contains eight hundred pounds of nitrogen on the steeper slopes and one thousand pounds on the more gently undulating areas, ours contains five thousand pounds in the brown silt loam and eight thousand pounds in the heavier black clay loam. This means that our Illinois prairie soil contains from five to ten times as much humus, or organic matter, as your best upland soil. To supply this difference in humus would require the addition of from four hundred to eight hundred tons per acre of average farm manure, or the plowing under of one hundred to two hundred tons of air-dry clover. This represents the great reserve of the Illinois prairie soils above the total supplies remaining in your soils.

"Our farmers are still producing crops very largely by drawing on this reserve. Of course most of this great supply of humus is very old. It represents the organic residues most resistant to decomposition; and, where corn and oats are grown exclusively, the soil has reached a condition on many farms under which the decomposition of the reserve organic matter is so slow that the nitrogen liberated from its own decay and the minerals liberated from the soil by the action of the decomposition products are not sufficient to meet the requirements of large crops, and for this reason alone some of our lands that are still rich are said to be run down; but they only require a moderate use of clover or farm manure or other fresh and active organic matter to at once restore their productiveness to a point almost equal to the yields from the virgin soil. Some Illinois farmers who have discovered this apparent restoration have jumped to the conclusion that they have solved the problem of permanently maintaining the fertility of the soil; and I judge from a remark made by the Secretary of Agriculture that some Iowa farmers have the same mistaken notions.

"These fresh supplies of active organic matter serve primarily as soil stimulants, hastening the liberation of nitrogen from the organic reserve and of minerals from the inorganic soil materials.

"Where one of the Eastern farmers has managed a farm under the rotation system with the occasional use of clover or light applications of farm manure,—where this has been continued until the great reserve is largely gone, and the phosphorus supply greatly depleted, then the land is truly run down, but not until then.

"Finally, land-plaster and quick-lime, still more powerful soil stimulants, are often brought into the system to bring about a more complete exhaustion of the soil reserves, and lastly the use of small amounts of high-priced commercial fertilizers serves to put the land in suitable condition for ultimate abandonment."

"Do you mean that commercial fertilizers injure the soil?" asked Mr. West.

"Well, to some extent they injure the soil because they tend to destroy the limestone and increase the acidity of the soil, and also because they contain more or less manufactured land-plaster and thus serve as soil stimulants; but the chief point to keep in mind concerning the use of the common so-called complete commercial fertilizer is that they are too expensive to permit their use in sufficient quantities to positively enrich the soil. Thus the farmer may apply two hundred pounds of such a fertilizer at a cost of $3.00 an acre, and then harvest a crop of wheat, two crops of hay, pasture for another year or two, plow up the grounds for corn, apply another two hundred pounds for the corn crop, follow with a crop of oats, and then repeat. He thus harvests five crops and pastures a year or two and applies perhaps four hundred pounds of fertilizer at a cost of $6.00.

"As an average of the most common commercial fertilizers sold to the farmers in the Eastern and Southern States, the four hundred pounds would add to the soil seven pounds of nitrogen, fourteen pounds of phosphorus and seven pounds of potassium, while a single fifty-bushel crop of corn will remove from the soil ten times as much nitrogen, five times as much potassium, and nearly as much phosphorus as the total amounts applied in this six-year or seven-year rotation.

"In this manner the farmer extends the time during which he can take from the soil crops whose value exceed their cost. He applies only one-fourth or possibly one-half as much of the most deficient element as the crops harvested require, and thus he continues for a longer time to 'work the land for all that's in it! '"

"Well, isn't that the limit?" said Adelaide, with emphasis on the "isn't," for which she received a disapproving look from her mother, so far as her almost angel-face could give such a look.

"So far as human ingenuity has yet devised," replied Percy, "this system appears to be the limit; but this limit has not yet been reached on any Westover soil. If anyone can devise a method for extending this limit he should apply it on a type of soil covering more than two-fifths of the total area of St. Mary County and more than 45,000 acres of Prince George County, Maryland, some of which almost adjoins the District of Columbia. This soil has been reduced in fertility until it contains only one-third as much phosphorus as your poorest land. I found a Western man who had come down to Maryland a few years ago. He saw that beautiful almost level upland soil, and it looked so good to him that he bought and kept buying until he had 'squared out' a tract of eleven hundred acres. He still had left money enough to fence the farm and to put the buildings in good repair. He was a live-stock farmer from the West who just knew from his own experience and from that of the Secretary of Agriculture, in the use of a little clover or farm manure in unlocking the great reserves of an almost virgin soil, that all his Maryland farm needed was clover seed and live stock. Sheep especially he knew to be great producers of fertility.

"He sowed the clover and grass seed and they germinated well. He even secured a fine catch, but it failed to hold, as we say out West. He tried again and again, and failed as often as he tried. He showed me his best clover on a field that had received some manure made from feed part of which was purchased, and that had also received five hundred pounds per acre of hydrated lime, which he was finally persuaded to use, after becoming convinced that clover-growing on old abandoned land was not exactly as easy as clover-growing on a 'run-down' farm of almost virgin soil in the West."

"And was the clover good after that treatment?" asked Mr. West.

"No, not good," said Percy, "but in some places where the manure had been applied to the high points, as is the custom of the Western farmer, the yield of clover, weeds, and foul grass together must have been nearly a half ton to the acre. Fortunately he waited to fully stock his farm with cattle and sheep until he should have some assurance of producing sufficient feed to keep them for a time at least, instead of making the common mistake of the less experienced farmer who goes to the country from the city, and who imagines that, if he has plenty of stock on the farm, they must of necessity produce abundance of manure with which to enrich his land for the production of abundant crops."

"Well, now you'll have to show me," said the grandmother. "To my way of thinking that's a pretty good kind of a notion for a farmer to have, and I'd like to know what's wrong with it."

Again a shadow seemed to cross the sweet face as the mother's glance turned from grandma to Adelaide.

"The system has some merit," replied Percy, "but it starts at the wrong point in the circle. Cattle and sheep must first have feed before they can produce the fertilizer with which to enrich the soil; and people who would raise stock on poor land should always produce a good supply of food before they procure the stock requiring to be fed. There is probably no more direct route to financial disaster than for one to insist upon over-stocking a farm that is essentially worn out."

"But doesn't pasturing enrich the soil?" asked the grandmother.

"Pasturing may enrich the soil only in a single element of plant food," said Percy. "In all other elements simple pasturing must always contribute toward soil depletion. If the pasture herbage contains a sufficient proportion of legume plants so that the fixation of free nitrogen exceeds the utilization of nitrogen in animal growth, then the soil will be enriched in that element, although with the same growth of plants it would be enriched more rapidly without pasturing; for animals are not made out of nothing. Meat, milk, and wool are all highly nitrogenous products.

"On the other hand no amount of pasturing can add to the soil a single pound of any one of the six mineral elements, and phosphorus, which is normally the most limited of all these elements, is abstracted from the soil and retained by the animals in very considerable amounts. As an average one-fourth of the phosphorus contained in the food consumed is retained in the animal products, especially in bone, flesh, and milk."

"Well, I didn't know that milk contained phosphorus," said Mr. West, "although I did know, of course, that phosphorus must be contained in bone."

"But, as you know," said Percy, "milk is the only food of young animals, and they must secure their bone food from the milk. Furthermore, the complete analysis of milk shows that it contains very considerable quantities. There are also records of digestion experiments in which less than one-half of the phosphorus in the food consumed was recovered in the total manural excrements. As a matter of fact there is a time in the life of the young mother, as with the two-year old cow, for example, when she must abstract from the food she consumes sufficient phosphorus for the nourishment of three growing animals,—her own immature body, a suckling calf, and another calf as yet unborn.

"Of course the organic matter of the soil should increase under pasturing, especially under conditions that make possible an accumulation of nitrogen; but here too the animals make no contribution toward any such accumulation. With the same growth of plants the accumulation of organic matter would be much more rapid without live stock."

"It is known absolutely but not generally that live stock destroy about two-thirds of the organic matter contained in the food they consume. With grains the proportion is higher, and with coarse forage it is lower, but as an average about two-thirds of the dry matter in tender young grass or clover or in a mixed, well-balanced ration of grain and hay is digested and thus practically destroyed so far as the production of organic matter is concerned.

"This you could easily verify yourself, Mr. West, by feeding two thousand pounds of any suitable ration, such as corn and clover hay, collecting and drying the total excrement, which will be found to weigh about seven hundred pounds, if it contains no higher percentage of moisture than was contained in the two thousand pounds of food consumed.

"Of course one should not forget that the liquid excrement contains more nitrogen and more potassium than the solid, and that much of this can be saved and returned to the land by use of plenty of absorbent bedding, and in pasturing there is no danger of any loss from this source."

"That is one great trouble with us," said Mr. West. "We never have as much bedding as we could use to advantage, and it is altogether too expensive to permit us to think of buying straw."

"Probably it would be much less expensive for you to buy ground limestone and then use good alfalfa hay for bedding," said Percy. "I mean exactly what I say," he continued. "Of course I do not advise you to use good alfalfa hay in that way, but it would be a cheap source of very valuable bedding, and it would make an extremely valuable manure. However, I should not hesitate to make liberal use of partially spoiled alfalfa hay for bedding, and you are quite likely to have more or less such hay; for under favorable conditions, such as you can easily have with your soil and climate, alfalfa comes on with a rush in the spring, and often the first crop should be cut before the weather is suitable for making hay. There should be very little or no delay at this time, because the first cutting should be removed in order that it may be out of the way of the second crop, which comes forward still more rapidly under normal conditions.

"Some of our Illinois farmers make strenuous objection to taking care of an alfalfa field that produces $50 worth of the richest and most valuable hay, because it interferes too much with the proper care of a $25 corn crop, which they somehow feel requires and deserves all their time and attention.

"Some of our Virginia farmers have sent to Illinois for their seed corn," said Mr. West; "and they report very good results as a rule, especially on land that has been kept up. On our poor land I think the native corn does better than the Western seed."

"Perhaps that is because it is used to it," suggested Percy, "used to making the struggle for itself on poor land. Fighting for all it gets, so to speak. You know the high-bred animals cannot hold their own with the scrubs when it comes to pawing the snow off the dead wild grass for a living in the winter, as cattle must do sometimes on the plains of the Northwest.

"Well, there may be something in that," responded Mr. West, "but the western seed corn certainly looks fine."

"Yes, that is true," said Percy. "Our farmers have made marked improvement in seed corn; they also understand very well how to grow corn. They know how and when to prepare the ground, how and when to plant; and how and when to cultivate. When Illinois farmers go to Iowa to buy land, the Iowa real estate men usually take them to see a farm that is owned and operated by a former Illinoisan, and they insist that there are no other farmers who know how to raise corn quite so well as the Illinois farmer. Perhaps the Illinois real estate man would tell a similar story to the Iowa farmer if he ever came there to buy land, but 'Westward the Course of Empire takes its Way' and the man once gone west knows the east no more, except as a market for his surplus products or a good place in which to spend his surplus cash.

"But, here. We must finish our study of the data that Miss Adelaide so kindly helped me to compute."

It was the first time that he had spoken her name in her presence; and she met his glance as she raised her eyes.

What's in a name? What's in a glance?

Percy proceeded without delay; and Adelaide listened as before, her drooping lashes protecting her eyes almost entirely from the view of others. The father and mother heard no name spoken and saw no eyes meet, and yet as Percy continued speaking a second self seemed to be thinking different thoughts and he was conscious of a strong desire to look longer than an instant into those captivating eyes.

A side glance, as she let her lashes droop, revealed to Adelaide that grandma alone had heard and seen. But Percy was a very common-place man. Certainly he had no such face as had held her glance for more than an instant as the afternoon train began to move from the depot platform. Percy was slightly above the average height and solidly built, but he was not tall. His face had often been described as a "perfect blank." No one saw anything of what lay within by merely looking into his eyes, and yet there was a certain indescribable something that appealed to one from those eyes. An elderly German lady once remarked to his mother: "Ihr Sohn hat so etwas gutes im Auge."

Percy was not polished in manner, Adelaide admitted. Professor Barstow had said that he deliberated for half an hour as to whether he should bring his "cawds," for use on Thanksgiving day, because he feared that the custom in "Vi'ginia" might not be the same as in "No'th Cahlina"; while she doubted very much if Percy had any cards whatever. She had never heard it said that he was "strong as an ox and quick as lightning," but perhaps she knew it as well as his schoolmates ever had. She had not heard that one of the college professors, noted for his short-cut expressions, had once told his class that he wished they would all "keep their thinking apparatus in as good repair as Johnston's." One thing she did know was that Percy's voice had been trained to talk to a woman, and that no other voice had ever spoken her name as he did. Reserve force? depth of manhood? confidence in his own words? absolute decision? wealth of tenderness? persistent endurance? unfailing loyalty? boundless affection? Deep in her heart Adelaide felt that these were among the attributes revealed in Percy's voice. When he spoke all listened. His voice was low-pitched but rich in tone and volume and sincerity,—that was the word.—The whole man seemed to feel and speak when he spoke. He surely can have no secrets. His mother must know all that he knows of his own self; but were those letters from his mother? The handwriting was very modern. Even her father made an old-fashioned C and W in signing his own name. Had he not looked at the writing on both those letters before he noticed the others? and why did he remain so long in his room before coming down to dinner? Had he not been in college—in a great University where there were hundreds of the brightest girls of his own State? But why should any girl be interested in farming? Teaching is such a cultured profession.

Only a moment—just while he was sorting the papers upon which they had made the computations, but a hundred thoughts had passed through her mind. Now he was speaking.

"You remember we took a sample of the subsoil on the sloping land. This soil is evidently residual, formed in place from the disintegration of the underlying rock. The soil may represent only a small part of the original rock, because of the loss by leaching. Here are the amounts of plant food found in two million pounds of the subsoil:

590 pounds of nitrogen 1,980 pounds of phosphorus 37,940 pounds of potassium 24,808 pounds of magnesium 31,320 pounds of calcium

"A splendid subsoil," Percy continued. "I know of none better in Illinois, except that we sometimes have more calcium in the form of carbonate, and even somewhat more potassium in places; but this must be a fine subsoil for alfalfa, where the bed rock is not too near the surface. Of course there is but little nitrogen in the subsoil, but that is true of all normal soils, because the nitrogen is contained only in the organic matter, and that decreases rapidly with depth and usually becomes insufficient to color the soil below 18 inches."

"Now," began Mr. West, "from these different analyses or invoices, and from your discussion of these results, I take it that you would not advise me to purchase any commercial fertilizer for use on the land we are still using in my rotation; but you think we should make large use of limestone and legume crops."

"Yes, Sir. Phosphorus is markedly deficient only in the very level upland which has been allowed to remain uncleared for fifty years or more, and nitrogen is certainly the limiting element on the land you are trying to keep in your rotation. While you cannot hope to put into your soil any such reserve of slow-acting organic matter as we still have in our comparatively new soils of the West, we may keep in mind that a small amount of quick-acting fresh organic matter is more effective than a large supply of what we might call embalmed material that decomposes very, very slowly unless assisted by the addition of more active organic matter. It frequently happens that one soil containing a large reserve of old humus, and hence showing more organic carbon and more nitrogen, by the ultimate invoice, than another soil, is, nevertheless, less productive, because the other soil contains a larger amount of fresh organic matter which decays quickly and thus furnishes more nitrogen and liberates more of the other elements from the insoluble minerals of the soil because of the greater abundance of the active products of organic decay.

"I think you should keep in mind, however, that, for every twenty-five bushels of corn you wish to produce, you should return to the soil one ton of clover or four tons of average farm manure, and that for one ton of produce hauled to the barns and fed, you will probably not return to the land more than one ton of manure."

CHAPTER XXX

"STONE SOUP"

THE next forenoon Percy and Mr. West spent some time making some further tests with hydrochloric acid and litmus paper in different places on the farm; but the result only confirmed the previous examinations.

"I never before saw any such light as now appears," said Mr. West. "It seems to me that for the first time in the history of Westover, covering about two centuries, a real plan can be intelligently made based upon definite information looking toward the positive improvement of the soil. While you have been away, I have been looking up the lime matter. I find that a lime is being advertised, and sold in small amounts, that is called hydrated lime, and it is especially prepared as an agricultural lime. It is recommended by some dealers as being fully equal to the ordinary commercial fertilizer which sells at about $25 a ton, while this hydrated agricultural lime can be bought for $8 a ton, and I think for a little less in larger amounts. You mentioned also that you had seen some one who had used hydrated lime, but it didn't seem to make much of a clover crop. Of course, I understand from what you said that his soil contained only one hundred and sixty pounds of phosphorus, and I take it that lime alone could not markedly improve his soil; but still I would like to know why, if he has one hundred and sixty pounds of phosphorus in his plowed soil, he could not produce a few good crops of clover. HOW much phosphorus does it require for a ton of clover?"

"One ton of clover contains only five pounds of phosphorus," Percy replied, "and of course the roots must also require some phosphorus, although after the crop is produced and removed, the phosphorus contained in the roots remains for the benefit of subsequent crops. Thus we might suppose the land which contains one hundred and sixty pounds of phosphorus ought to furnish the phosphorus needed for a three ton crop of clover every year for ten years; but in actual practice no such results are secured. The invoice of the plant food in the soil is a matter of very great importance, for it reveals the mathematical possibilities, but another matter of almost equal importance is the problem of liberating plant food from this supply sufficient for the crops to be produced year by year.

"Decaying or active organic matter is one of the great factors in the liberation of plant food, and undoubtedly the extension or distribution of the root system of the growing plant is another very potent factor. If the root surfaces come in contact with one per cent. of the total surface of the soil particles in the plowed soil, then we might conceive of a relationship whereby one per cent. of the phosphorus in that soil would be dissolved or liberated from the insoluble minerals and thus become available as food for the growing crop. We know that the rate of liberation varies greatly, with different soils and seasons, and crops also differ in their power to assist themselves in the extraction of mineral plant food from the soil. The presence of limestone encourages the development of certain soil organisms which tend to hasten some decomposition process. But, all things considered, it may be said, speaking very generally, that the equivalent of about one per cent. of the total phosphorus contained in the plowed soil does become available for the crops under average conditions. On this basis one hundred and sixty pounds of phosphorus would furnish about one and one-half pounds for the crops during one season. But in such a soil the phosphorus still remaining may be the most difficultly soluble, and the supply of decaying organic matter may be extremely low, so that possibly less than one pound per acre would become available, and this would meet the needs of less than four hundred pounds per acre of clover hay. Furthermore, the supply grows less and less with every crop removed.

"With your ordinary soil, carrying twelve hundred and seventy pounds of phosphorus, perhaps you may be able by a liberal use of decaying organic matter to liberate ten or fifteen pounds of phosphorus, or sufficient for a crop of forty to sixty bushels of corn; and, with a subsoil richer in phosphorus than the surface, and with more or less of the partially depleted surface removed by erosion year by year, the supply of phosphorus is thus permanently provided for unless the bed rock is brought too near the surface. It is doubtful if the direct addition of phosphorus to your sloping lands will ever be necessary or profitable. Certainly such addition is not advisable until you have brought the land to as high a state of fertility as is practicable by means of limestone, legumes, and manure."

"That seems clearly to be the case with most of the land now under cultivation on this farm," said Mr. West "Can you tell me anything about this hydrated lime?

"I can tell you it is correctly named," Percy replied. "*Hydrated* means *watered,* and an investment in hydrated lime is properly classed with other watered investments. If you prefer to use hydrated lime I would suggest that you buy fresh burned lump lime and do the hydrating yourself, which only requires that you add eighteen pounds of water to each fifty-six pounds of quick lime; in other words, that you slack the lime by adding water in the proper proportion. Both quick lime and hydrated lime are known as caustic lime. Webster says that the word *caustic* means 'capable of destroying the texture of anything or eating away its substance by chemical action.'

"This definition is correct for caustic lime, as you can easily determine by keeping your hand in a bucket of slacked lime a few minutes. Caustic lime eats away the organic matter of the soil. In an experiment conducted by the Pennsylvania Experiment Station, during a period of sixteen years, eight tons of hydrated lime destroyed organic matter equivalent to thirty-seven tons of farm manure, as compared with the use of equivalent applications of ground limestone; and, as an average of the sixteen years, every ton of caustic lime applied liberated seven dollars' worth of organic nitrogen, as compared with ground limestone. That this much liberated nitrogen was essentially wasted and lost is evidenced by the fact that larger crops were produced where ground limestone was used than where burned lime was applied.

"The limestone must be quarried whether used for grinding or for burning, and the grinding can be done for twenty-five cents a ton where a large equipment with powerful machinery is used and where cheap fuel is provided, as near the coal mining districts. It need not be very finely ground. If ground to pass a sieve with twelve meshes to the linear inch, it is very satisfactory, provided that all of the fine dust produced in the grinding is included in the product. You see the soil acids are slightly soluble and they attack the limestone particles and are thus themselves destroyed or neutralized. If, however, you ever wish to use raw rock phosphate, insist upon its being sufficiently fine-ground that at least ninety per cent. of it will pass through a sieve with ten thousand meshes to the square inch, this being no finer than is required for the basic slag phosphate, of which several million tons are now being used each year in the European countries. Like the raw rock phosphate, the slag gives the best results only when used in connection with plenty of decaying organic matter."

"That reminds me," said Mr. West, "of what one of the fertilizer agents said about raw phosphate. He said the use of raw phosphate with farm manure reminded him of 'stone soup,' which was made by putting a clean round stone in the kettle with some water. Pepper and salt were added, then some potatoes and other vegetables, a piece of butter and a few scraps of meat. 'Stone soup,' thus made, was a very satisfactory soup. He said that in practically all of the tests of raw phosphate conducted by the various State Experiment Stations, manure has been used as a means of supplying organic matter to liberate the phosphorus from the raw rock, but in such large quantity as to be entirely impracticable for the average farmer to use on his own fields; and his opinion was that the entire benefit was due to the manure. He had a little booklet entitled 'Available or Unavailable Plant Food—Which?' published by the National Fertilizer Association, and said I could get a copy by addressing the Secretary at Nashville, Tennessee."

"Fortunately," said Percy, "this is not a question of opinion but one of fact; and it has been discovered that the fertilizer agents who are long on opinions and short on facts prefer to sell four tons of complete fertilizer for $80, or even two tons of acid phosphate for $30, rather than to sell one ton of raw phosphate, containing the same amount of phosphorus, for $7.50. In the manufacture of acidulated fertilizers, one ton of raw phosphate, containing about two hundred and fifty pounds of the element phosphorus, is mixed with one ton of sulfuric acid to make two tons of acid phosphate; and, as a rule, these two tons of acid phosphate are mixed with two tons of filler to make four tons of complete fertilizer. A favorite filler is dried peat, which is taken from some of the peat bogs, as at Manito, Illinois, and shipped in train loads to the fertilizer factories. The peat is not considered worth hauling onto the land in Illinois, even where the farmers can get it for nothing; but it contains some organic nitrogen, and, by the addition of a little potassium salt, the agent is enabled to call the product a 'complete' fertilizer.

"Experiments with the use of raw rock phosphate have been conducted by the State Agricultural Experiment Stations over periods of twelve years in Maryland, eleven years in Rhode Island, twenty-one years (in two series) in Massachusetts, fourteen years (in two series) in Maine, twelve years in Pennsylvania, thirteen years in Ohio, four years in Indiana, and from four to six years on a dozen different experiment fields in different parts of Illinois.

"I have here some quotations taken from the directors of several of these experiment stations which fairly represent the opinions which they have expressed concerning their own investigations. Thus the Maryland director says:

"'The results obtained with the insoluble phosphates has cost usually less than one-half as much as that with the soluble phosphates. Insoluble South Carolina phosphate rock produced a higher total average yield than dissolved South Carolina rock.'

"The Rhode Island director comments as follows:

"' With the pea, oat, summer squash, crimson clover, Japanese millet, golden millet, white podded Adzuka bean, soy bean, and potato, raw phosphate gave very good results; but with the flat turnip, table beet, and cabbage it was relatively very inefficient.'

"The following statement is from the Massachusetts director:

"'It is possible to produce profitable crops of most kinds by liberal use of natural phosphates, and in a long series of years there might be a considerable money saving in depending at least in part upon these rather than upon the higher priced dissolved phosphates.'

"The director of the Maine State Experiment Station gives us the following:

"'For the first year the largest increase of crop was produced by soluble phosphate. For the second and third years without further addition of fertilizers, better results were obtained from the plots where stable manure and insoluble phosphates had been used.'

"The stable manure and insoluble phosphates here referred to were not applied together, but on separate plots. In deed, the raw phosphate was not used in connection with manure either in Maryland, Rhode Island, Massachusetts, Maine, Pennsylvania, or Indiana; and in the extensive experiments in progress in Illinois the raw phosphate has been used, as a rule, not with farm manure, but with green manures; and wherever manure has been used in connection with the raw phosphate, as in Ohio, the comparison is made with the same amounts of manure applied without phosphate.

"The Pennsylvania Report for 1895, page 210, contains the following statement:

"'The yearly average for the twelve years gives us a gain per acre of $2.83 from insoluble ground bone, $2.45 from insoluble South Caroline rock, $1.61 from reverted phosphate, and 48 cents from soluble phosphate, thus giving us considerably better results from the two forms of insoluble phosphate than from the reverted or soluble forms.'

"The Indiana director reports as follows:

"'It will be seen that during the first and second years the rock phosphate produced little effect, while the acid phosphate very materially increased the yields. During the third and fourth seasons, however, the rock produced very striking results, even forging ahead of the acid. This and very similar investigations in progress lead us to believe that rock phosphate is a cheap and effective source of phosphorus where immediate returns are not required.

"In the Ohio experiments eight tons of manure per acre were applied once every three years in a three-year rotation of corn, wheat, and clover, three different fields being used, so that every crop might be grown every year. The average yields for the thirteen years where manure alone was used were:

53.1 bushels of corn 20.6 bushels of wheat 1.63 tons of hay

"The average yields on the unfertilized land were:

32.2 bushels of corn 11.4 bushels of wheat 1.16 tons of hay

"If the corn is worth 35 cents a bushel, the wheat 70 cents, and the hay $6 a ton, in addition to the expense of harvesting and marketing, then the total value of the manure spread on the land is $2.07 a ton.

"Where $1.20 worth of raw phosphate (320 pounds) were added in connection with the manure the average yields were as follows:

61.4 bushels of corn 26.3 bushels of wheat 2.23 tons of hay

"And where $2.40 worth of acid phosphate (320 pounds) were used with the same amount and kind of manure the following average yields were secured:

60.4 bushels of corn 26.5 bushels of wheat 2.16 tons of hay

"These are the actual yield, and by any method of computation yet proposed, each dollar invested in raw phosphate has paid back much more than has a dollar invested in acid phosphate."

"And was the use of the raw phosphate really profitable?" asked Mr. West.

"Well, you might figure that out for yourself," Percy replied, "preferably using the average prices for your own locality for corn, wheat and clover. As I figure it at prices below the ten-year average for Illinois, the raw phosphate paid about eight hundred per cent. net on the investment."

"Eight hundred per cent! You must mean eight per cent. net.

"No, Sir, I mean eight hundred per cent. net, but you had better take the data and make your own computations. But does it not seem strange that, with such positive knowledge as this available, many of the Illinois landowners who have managed to sell off enough of their original stock of fertility in grain or stock at good prices to enable them to more than pay for their lands, should continue to invest their surplus in more land with hope that it will pay them eight per cent. interest, when they could secure many times that much interest from investing in the permanent improvement of the land they already own?"

"Perhaps it is not so strange," replied Mr. West. "I fear that some of their ancestors did the same thing in Virginia and other Eastern States until the land became poor, and then of course they were 'land poor.' But, say, that 'stone soup' wouldn't be so bad for those Ohio landowners, would it? I should think they would avail themselves of the positive information from their experiment station. Speaking of soup, I wonder if it isn't time for lunch! But tell me; are the Illinois farmers doing anything with raw phosphate?"

"Yes, they are doing something, but by no means as much as they ought. About two months ago a group of the leading farmers from our section of the State went up to Urbana to look over the experiment fields, some of which have been carried on since 1870. The land is the typical corn belt prairie, and consequently the results should be of very wide application. Well, as a result of that day's inspection of the actual field results, an even twelve carloads of raw phosphate were ordered by those farmers upon their return home; and I learned of another community where ten carloads were ordered at once after a similar visit. As an average of the last three years the yield of corn on those old fields has been 23 bushels per acre where corn has been grown every year without fertilizing, 58 bushels where a three-year rotation of corn, oats and clover is followed, and in the same rotation where organic matter, limestone, and phosphorus have been applied the average yield has been 87 bushels in grain farming and 92 bushels in live-stock farming.

"I attended the State Farmers' Institute last February, and there I met many men who have had several years' experience with the raw rock. Usually they put on one ton per acre as an initial application and plow it under with a good growth of clover; and, afterward, about one thousand pounds per acre every four years will be ample to gradually increase the absolute total supply of phosphorus in the soil, even though large crops are removed.

"A good many of our thinking farmers are now using one or two cars of raw phosphate every year, and they are figuring hard to keep up the organic matter and nitrogen. The most encouraging thing is the very marked benefit of the phosphate to the clover crop, and of course more clover means more corn in grain farming, and more corn and clover means more manure in live-stock farming.

"On the Illinois fields advantage is taken of these relations in the developing of systems of permanent agriculture. You see, if the phosphate produces more clover, then more clover can be plowed under on that land; or, if the crops are fed, then more manure can be returned to the phosphated land than to the land not treated with phosphate and not producing so large crops. Really the phosphate is not given full credit for what it has accomplished in the Ohio experiments; because, while the land receiving phosphated manure has produced about one-fourth larger crops than the land receiving the untreated manure, the actual amounts of manure applied have been the same, whereas one-fourth more manure can be produced from the phosphated land and if this increased supply of manure were returned to the land it would increase the supply of nitrogen and thus make still larger crop yields possible."

"That is surely the way it would work out in practical farming," said Mr. West. "I think I did not tell that $4.80 a ton is the lowest quotation I have been able to get as yet for ground limestone delivered at Blue Mound Station."

"That would make its use prohibitive," said Percy. "You ought to get it for just one-fourth of that, or for $1.20 a ton. In Illinois we can get it delivered a hundred miles from the quarry for $1.20 a ton. It costs no more for a thirty-ton car of ground limestone than the farmer receives for a cow; and the cost of a car of fine-ground natural phosphate is about equal to the price of one horse."

"Of course, our limestone supplies are essentially inexhaustible," said Mr. West, "but is that also true of our natural phosphate deposits?"

"It is not true of the high-grade phosphate," replied Percy; "for, according to the information furnished by the United States Geological Survey, it is evident that the known supplies of our high-grade phosphate will be practically exhausted in fifty years if our exportation continues to increase at the prevailing rate. After that is gone we may then draw upon our low-grade phosphate deposits, which though probably not inexhaustible are known to be exceedingly extensive."

CHAPTER XXXI

THEORIES VERSUS FACTS

PERCY planned to walk to Blue Mound to take the three-thirty train that Saturday afternoon; but Adelaide's parents both insisted that she would willingly drive to the station, and the grandmother discovered that she needed a certain kind of thread which Adelaide could then also get at the store.

"Certainly," said Adelaide, with some merriment, "I always enjoy our departing guests to the train."

"Very well," replied Percy. "If you must go to get the thread and will permit me to be the coachman, I shall be satisfied, for you will be home early."

"Then we will take the colts and buckboard, and I shall be home in less than twenty minutes after your train leaves the station."

"I think I have missed several days of your beautiful 'Indian Summer,' because of my trip to the North," Percy remarked to Mr. West as they sat on the broad veranda waiting for the hour of two thirty when the colts were to be ready for the drive.

"I wish you might have been with us while Professor Barstow was here," replied Mr. West, "not only because of the mild autumn weather we have had, but also because Professor Barstow has some ideas about questions of soil fertility that are very different from those you hold. He says a young man from Washington gave a lecture at his college down in North Carolina, in which he informed them that the cause of infertility of soils is a poisonous substance excreted by the plant itself, and that this can be overcome by changing from one crop to another because the excrete of one plant, while poisonous to that plant, will not be poisonous to other plants of a different kind. Thus, by rotation of crops, good crops could be grown indefinitely on the same land without the addition of plant food. He said that the soil water alone dissolved plenty of plant food from all soils for the production of good crops, and that the supply of plant food will be permanently maintained, because the plant food contained in the subsoil far below where the roots go is being brought to the surface by the rise of the capillary moisture, and that there is in fact a steady tendency toward an accumulation of plant food in the surface soil. He said that it is never necessary to apply fertilizing material to any soil for the purpose of increasing the supply of plant food in that soil. He admitted that applications of fertilizers sometimes produce increased crop yields, but that the effect was due to the power of the fertilizer to destroy the toxic substances excreted by the plants, and this is really the principal effect of potash, phosphates, and nitrates, and also of farm manure and green manures. Humus, he said, is one of the very best substances for destroying these toxic excrete although they had some other things which were as good or better than any sore of fertilizing materials. He mentioned especially a substance called pyrogallol, which cost $2.00 a pound, and of course it could not be applied on a large scale; but it was as good a fertilizer as anything, although it contains nothing but carbon, oxygen, and hydrogen, which, as you explained to me when you were here before, the plants secure in abundance from air and water. This information had been secured in the laboratories at Washington by growing wheat seedlings in water culture for twenty-day periods."

"I have already heard something of those theories," said Percy, "but I shall be glad to have you tell me more about them. As I understand them, we need only to rotate and cultivate and our lands should always continue to produce bountiful crops. Is that correct?"

"I understand that is the theory," replied Mr. West, "but I know it is not correct for my grandfather used to grow two or three times as much wheat per acre as I can grow, and I rotate much more than he did. In fact I can grow only ten to fifteen bushels of wheat per acre once in ten years, whereas he grew from twenty-five to forty bushels per acre in a five-year rotation; and I don't see that there is any particular connection between the growing of wheat seedlings in small pots or bottles for a few twenty-day periods and the growing of crops in soils during successive seasons. No, I don't take any stock in their theories. I think they are _watered, _or perhaps I should say _hydrated, _in deference to science. But I would like to know about this question of plant food coming up from below. That would be a happy solution of the fertilizer problem."

"It is true," said Percy, "that soluble salts are brought to the surface in the rise of moisture by capillarity in times of partial drouth; and in the arid regions where the small amount of water that falls in rain or snow leaves the soil only by evaporation, because there is never enough to produce underdrainage, the salts tend to accumulate at the surface. The alkali conditions in the arid or semiarid regions of the West are thus produced. But in humid sections where more or less of the rainfall leaves the soil as underdrainage the regular loss by leaching is so much in excess of the rise by capillarity that soils which are not affected by erosion or overflow steadily decrease in fertility even under natural conditions, with no cultivation and no removal of crops. Of course this applies at first only to the mineral plant foods, as phosphorus potassium, magnesium, and calcium. While mineral supplies are abundant in the surface soil, there may be a large acumulation of organic matter and nitrogen, especially because of the growth of wild legumes, which are very numerous and in places very abundant, especially on some of the virgin prairies of the West. However, as the process of leaching proceeds there comes a time when the growth of the native vegetation is limited because of a deficiency in some essential mineral plant food, such as phosphorus, or the limestone completely disappears and soil acidity develops which greatly lessens the growth of the legumes.

"Decomposition of organic matter begins almost as soon as any part of the plant ceases to live, and there is certain to come a time when the rate of decomposition and loss exceeds the rate of fixation and accumulation; and from that time on the organic matter and nitrogen as well as the mineral plant foods continue to decrease in the surface, until finally the natural barrens are developed, such as are found in different sections of the World and in some places even where the rainfall is sufficient for abundant crops."

"Yes, Sir," said Mr. West. "I know that is true. I have visited Tennessee and I know there are some extensive areas there of practically level upland which have always been considered too poor to justify putting under cultivation, and they are called the 'Barrens'."

"I know about those barren lands, too," said Percy. "Our teacher of soil fertility in college told us that a farm is more than a piece of the earth's surface. He said if we only wanted to get a large level tract of upland where the climate is mild and the rainfall abundant and where all sorts of crops do well on good soil, including the wonderful cotton crop which brings a hundred dollars for a thousand pounds, while corn brings forty dollars for a hundred bushels,—well, he said we could go to the Highland Rim of Tennessee where, according to analyses reported in 1897 by the Tennessee Experiment Station, the surface soil of the 'Barrens' contains eighty-seven pounds of phosphorus and the subsoil sixty-one pounds of phosphorus per acre, counting two million pounds of soil in each case. He said, if we didn't like that we might go into the Great Central Basin of Tennessee or the famous Blue Grass Region of Kentucky and find land that is still extremely productive and more valuable than ever, even after a hundred years of cultivation, and buy land containing from three thousand to fifteen thousand pounds of phosphorus per acre."

"I know both of those sections very well," said Mr. West. "But doesn't it seem strange that the scientists at Washington would teach as they do? Why doesn't the plant food accumulate in the surface soil of those barrens? Surely they have been lying there long enough, with no crops whatever removed, so that they ought to have become very rich. I wish I had known about their phosphorus content so I could have told Professor Barstow. He was quite carried away with the Washington theory."

"You ought not for a moment call it the 'Washington' theory," said Percy; "and neither is it promulgated by scientists, but rather by two or three theorists who are upheld by our greatest living optimist. Science, Sir, is a word to be spoken of always with the greatest respect. Of course you know its meaning?"

"Yes, I know it comes from the Latin _scire, _to know."

"Then _science _means _knowledge; _it does not mean theory or hypothesis, but absolute and positive knowledge. Is there any uncertainty as to the instant when the next eclipse will appear? No, none whatever. Science means knowledge, and men are scientists only so far as they have absolute knowledge, and to that extent every farmer is a scientist.

"Nevertheless the erroneous teaching so widely promulgated by the federal Bureau of Soils is undoubtedly a most potent influence against the adoption of systems of positive soil improvement in the United States, because it is disseminated from the position of highest authority. Other peoples have ruined other lands, but in no other country has the powerful factor of government influence ever been used to encourage the farmers to ruin their lands."

CHAPTER XXXII

GUESSING AND GASSING

AS we were riding to Montplain yesterday," said Adelaide to Percy, soon after they started for Blue Mound, "Professor Barstow told me that in his opinion all that was needed to redeem these old lands of Virginia and the Carolinas is plenty of efficient labor, such as he thinks we had before the war. I know papa does not agree with him in that, but Professor said that soils do not wear out if well cultivated, that in New England they grow as large crops as ever, and that in Europe, on the oldest lands the crop yields are very much larger than in the United States; and in fact that the European countries are producing about twice as large crops as they did a hundred years ago. He thinks it is because they do their work more thoroughly than we do. He says that 'a little farm well tilled' is the key to the solution of our difficulties."

"That might seem to be a good guess as to the probable relation of cause and effect," replied Percy, "but we ought not to overlook some well known facts that have an important bearing. It is exactly a hundred years since DeSaussure of France, first gave to the world a clear and correct and almost complete statement concerning the requirements of plants for plant food and the natural sources of supply. Sir Humphrey Davy, Baron von Liebig, Lawes and Gilbert, and Hellriegel followed DeSaussure and completely filled the nineteenth century with accumulated scientific facts relating to soils and plant growth.

"Sir John Bennett Lawes, the founder of the Rothamsted Experiment Station, the oldest in the world, on his own private estate at Harpenden, England, began his investigations in the interest of practical agricultural science soon after coming into possession of Rothamsted in 1834. In 1843 he associated with him in the work Doctor Joseph Henry Gilbert, and for fifty-seven years those two great men labored together gathering agricultural facts. Sir John died in 1900, and Sir Henry the following year.

"That the people of Europe have made some use of the science thus evolved is evident from the simple fact that they are taking out of the United States every year about a million tons of our best phosphate rock for which they pay us at the point of shipment about five millions dollars; whereas, if this same phosphate were applied to our own soils that already suffer for want of phosphorus, it would make possible the production of nearly a billion dollars' worth of corn above what these soils can ever produce without the addition of phosphorus. And our phosphate is only a part of the phosphate imported into Europe. They also produce rock phosphate from European mines, and great quantities of slag phosphate from their phosphatic iron ores.

"They feed their own crops and large amounts of imported food stuffs, and utilize all fertilizing materials thus provided for the improvement of their own lands. Legume crops are grown in great abundance and are often plowed under to help the land.

"Do you wonder why the wheat yield in England is more than thirty bushels per acre while that of the United States is less than fourteen bushels? Because England produces only fifty million bushels of wheat, while she imports two hundred million bushels of wheat, one hundred million bushels of corn, nearly a billion pounds of oil cake, and other food stuffs, from which large quantities of manure are made; and, in addition to this, England imports and uses much phosphate and other commercial plant food materials.

"Germany imports great quantities of wheat, corn, oil cake, and phosphates, and thus enriches her cultivated soil, and Germany's principal export is two billion pounds of sugar, which contains no plant food of value, but only carbon, oxygen, and hydrogen, secured from air and water by the sugar beet.

"Denmark produces four million bushels of wheat, imports five million bushels of wheat, fifteen million bushels of corn, fifteen million bushels of barley, eight hundred million pounds of oil cake, eight hundred million pounds of mill feed, and other food stuffs, phosphate, etc., and exports one hundred and seventy-five million pounds of butter, which contains no plant food of value, but sells for much more than these imports cost.

"Italy applies to her soils every year about a million tons of phosphates, which contain nearly twice as much phosphorus as is removed from the land in all the crops harvested and sold from the farms of Italy.

"The very good yields of the crops of New England are attributable to large use of fertilizing materials, in part made from food stuffs shipped in from the West; and the high development of certain lands of Europe and New England has been possible under the system followed only because the areas concerned are small. Thus, the average acreage of corn in Rhode Island and Connecticut is less than three townships, or less than one-tenth as much corn land in the two States as the area of single counties in the Illinois corn belt.

"Did you ever hear of the 'Egypt' we have out West, Miss West?"

"Out West, Miss West," she repeated. "That is too much repetition of the same word to make a good sentence. I like 'Miss Adelaide' better; I do get tired of hearing West and Westover over and over. Yes, I have heard of the 'Egypt' you have out West. Is it near Illinois?"

"Near Illinois? Why, Miss Adelaide, I am surprised that you should even know about the crop yields of Rhode Island and not know where 'Egypt' is. Let me inform you that 'Egypt' is in Illinois, and our 'Egypt' is a country as large as thirteen states the size of Rhode Island. Cairo is the Capital, and Alexandria, Thebes, and Joppa are all near by. Tamm and Buncombe, and Goreville and Omega are also among our promising cities of 'Egypt,' although you may not so easily associate them with the ancient world."

"Well I know where Cairo is," Adelaide replied, "but if your 'Egypt' is on the map you will have to show me. I know now that

'Egypt' is in Southern Illinois, but how do you separate 'Egypt' from the rest of the State?"

"We make no such separation," said Percy. "But to find 'Egypt' on the map, you need only take the State of Illinois and subtract therefrom all that part of the corn belt situated between the Mississippi River and the west line of Indiana. The southern point of 'Egypt' is at Cairo, the Capital, and it is bounded on the east, south, and west, by the Wabash, the Ohio, and the Mississippi; but the north line is not only imaginary, but it is movable. In fact it is always just a few miles farther south, but I think all 'Egyptians' will agree that a sand bar which is being formed below Cairo between the Ohio and the Mississippi is truly 'Egyptian ' territory. If you ever visit in the West do not fail to see 'Egypt.'

"I really hope I may, sometime," she replied. "We have relatives who claim to live in Tennessee, Kentucky, and Missouri, but I think possibly they may all be 'Egyptians,' from what you have told me about the vast area of that great fairy empire. I know I would dearly love to go there."

"'Egypt' is the wheat belt and the fruit belt of Illinois," Percy continued. "One of the grand old men of Illinois, Colonel N. B. Morrison, who was for years a trustee of the State University, used to be called upon for an address whenever he was present at Convocation. He always stated proudly that he lived in the 'Heart of Egypt.' He said the soil there was not so rich perhaps as in the corn belt, but that with plenty of hard work they were able to live and to produce the finest fruit and the greatest men in America. He said they had to work both the top and bottom of their soil, and he explained that they harvested wheat and apples from the top, and then went down about 600 feet and harvested ten thousand tons of coal to the acre, and still left enough to support the earth. I have heard him say that when he was born there was not a mile of railroad in the United States, and that he had during his own lifetime, witnessed the practical agricultural ruin of almost whole States. He used to plead for the University to send some of her scientific men to help them to solve the problem of restoring the fertility of their soils down in 'Egypt'; and I am glad to say that finally the State appropriated sufficient funds so that the Illinois Experiment Station is rapidly securing the exact information needed to make those Southern Illinois lands richer than they ever were.

"I spent several days in 'Egypt' last month and I am planning to make another trip down there next week before deciding definitely about purchasing our poor land farm. I am not sure but the land of 'Egypt' is as poor as we ought to try to build up considering our limited means."

"Oh, do you think so? But Papa's land is not so poor is it?"

"No, it is not so poor in mineral plant food on the sloping areas, but even there it is extremely poor in humus and nitrogen. However, I fear I could not enjoy farming in irregular patches of five or ten acres, and the level lands of Virginia and Maryland are so exceedingly poor, that much time and money and work will be required to put them on a paying basis. There would be no pleasure or satisfaction in merely robbing other farms to build up mine, as some of the prosperous truck farmers and dairymen are doing. I should want to practice a system of soil improvement of unlimited application so that it would not be a curse to the agricultural people, as is the case with the man who builds up his farm only at the expense of other farms.

"We have been speaking of the development of agriculture on the small tracts of cultivable land in the great manufacturing States of New England. But, if we would make a fair comparison with a State like Illinois, we should consider some great agricultural State, as Georgia, for example, which is also one of the original thirteen. Georgia is a larger State than Illinois, and Georgia cultivates as many acres of corn and cotton as we cultivate in corn. But Georgia land cannot be covered with fertilizer made from Illinois corn, nor even with seaweed and fish-scrap from the ocean. Her agriculture must be an independent agriculture, just as the agriculture of Russia, India, and China must be, just as the agriculture of Illinois must be, and as the agriculture of all the great agricultural States must be. What is the result to date? The average yield of corn in Georgia is down to 11 bushels per acre. This is not for half of one township, but the average of four million acres for the last ten years; and this in spite of the fact that Georgia out more for the common acidulated manufactured so-called complete commercial fertilizer than any other State."

"That is appalling," said Adelaide, "but still some larger countries are building up their lands, such as those of Europe."

"In large part by the same methods as the New England truckers and dairymen are following," he replied, "and in comparison with the area and resources of their colonies and of the other great new countries upon which they draw for food and fertilizer, they are fairly comparable with the New England States in this country. Even the Empire of Germany is only four-fifths as large as Texas. The only country of Europe at all comparable with the United States is Russia, and in that great country the average yield of wheat for the last twenty years is eight and one-fourth bushels per acre, even though, as a general practice, the land is allowed to lie fallow every third year. The average yield for the five famine years that have occurred during the twenty-year period was six and one-quarter bushels of wheat per acre."

"That is wretched," said Adelaide, "I know about the Russian famines for we have made contributions through our church for their relief, but that condition can surely never come to this great rich new country, can it?"

"It will come just as certainly as we allow our soil fertility to decrease and our population to increase. As a nation we have scarcely lifted a hand yet to stop the waste of fertility or to restore exhausted lands; practically every effort put forth by the Federal government along agricultural lines having been directed toward better seeds, control of injurious insects and fungous diseases, exploitation of new lands by drainage and irrigation, popularly called 'reclamation,' although applied only to rich virgin soils which can certainly be brought under cultivation at any future time either by the Government or by private enterprise. But why should not the Federal government make all necessary provisions to furnish ground limestone and phosphate rock at the actual cost of quarrying, grinding, and transporting, in order that farmers on these old depleted soils may be encouraged to adopt systems of soil improvement; or even compelled to adopt such systems, just as they are compelled to build school houses, bridges, and battleships?"

"Perhaps the Government would do this," said Adelaide, "if the Secretary of Agriculture would recommend it."

"I have heard of the '_big if,'" _Percy replied slowly, "but I am afraid this _if _will beat the record for bigness. His soil theorists continue to assure him that soils do not wear out, no matter how heavily cropped, if they are only rotated and cultivated; and to support their theories they have forsaken the data from the most carefully conducted and long continued scientific investigations, and indulged in a game of guessing that the increasing productiveness of a few small countries of Europe is not due to any necessary addition of plant food.

"But here is the depot, and I have taken almost an hour to drive three miles. If I had hurried, you might have been back home by this time. I am afraid I have been selfish in allowing the team to walk nearly all of the way, but they will at least be fresh for the home trip which you promised to make in less than twenty minutes, I remember. Now if you will hold the lines, I will run into the store to get the thread. I remember the kind; I often do such errands for mother."

"I will wait while you get your ticket and find out if the train is on time," said Adelaide, as Percy returned with the thread.

"At least fifty minutes late," he reported, "and the agent said he was glad of it for he would need about that time to make out such a long-jointed ticket as I want. I am rather glad too, for I can watch you to the turn in the road on the hill, which must be a mile or more, and I will time you. You can have six minutes to make that corner."

"You mean I can have six minutes to get out of sight," she suggested.

"I think you are out of sight," he ventured.

Adelaide reddened. "I shall have to tell mother what slang you use," she said.

"I hope you will," he retorted, "for I have watched her watch you and I am sure she will agree with me. But I do feel that I owe you a sincere apology for taking up the time we have had together with this long discussion of the things that are of such special interest to me. I have been alone with my mother so much and she is always so ready and so able, I may add, to discuss any sort of business matter that I fear I have been forgetful of your forbearance."

"But you really have not," Adelaide replied. "I keep books for papa, and I am very much interested in these social and economic questions which are so fundamental to the perpetuity of our State and National prosperity. I have been both entertained and instructed by these discussions; and I might say, honored, too, that you do not consider me too young and foolish to talk of serious subjects."

"I am sure it is kind of you to make good excuses for me. You have at any rate relieved my mind of some burden, but I am sure you are the only woman I have ever known, except my mother, who could endure discussions of this sort. I have so greatly enjoyed the few short visits I have had with you. I wish I might write to you and I shall be so much interested to learn what success your father has if he begins a system of soil improvement. Would it be presuming to hope that I might hear from you also?"

"I am papa's stenographer," she replied, "and perhaps he will dictate and I will write. We will be glad to hear of your safe return,—and you,—you might ask papa. Now, I shall soon be out of sight."

"Please don't," begged Percy. "It is still forty-five minutes 'at least,' before the train comes. Let me go a piece with you. I will leave my suit case here and with nothing to carry I shall easily walk a mile in twenty minutes. May I drive, please?"

"No, I will drive. I want to ask you another question, and I am afraid you would drive too fast.

"You mentioned some long-continued scientific investigations which I assumed referred to the yield of crops. What were they?"

"I meant such investigations as those at Rothamsted and also those conducted at Pennsylvania State College. I have some of the exact data here in my note book.

"In 1848, Sir John Lawes and Sir Henry Gilbert began at Rothamsted, England, two four-year rotations. One was turnips, barley, fallow, and wheat; and the other was turnips, barley, clover, and wheat. Whenever the clover failed, which has been frequent, beans were substituted, in order that a legume crop should be grown every fourth year.

"The average of the last twenty years represents the average yields about fifty years from the beginning of this rotation.

"In the legume system, as an average of the last twenty years, the use of mineral plant food has increased the yield of turnips from less than one-half ton to more than twelve tons; increased the yield of barley from thirteen and seven-tenths bushels to twenty-two and two-tenths bushels; increased the yield of clover (when grown) from less than one-half ton to almost two tons; increased the yield of beans (when grown) from sixteen bushels to twenty-eight and three-tenths bushels; and increased the yield of wheat from twenty-four and three-tenths bushels to thirty-eight and four-tenths bushels per acre.

"In the legume system the minerals applied have more than doubled the value of the crops produced, have paid their cost, and made a net profit of one hundred and forty per cent. on the investment, in direct comparison with the unfertilized land.

"If we compare the average yield of turnips, barley, clover, and wheat of the last twenty years with the yield of turnips in 1848, barley in 1849, clover in 1850 and wheat in 1851 we find that on the unfertilized land in this rotation of crops in fifty years the yield of turnips has decreased from ten tons to one-half ton, and the yield of barley has decreased from forty-six to fourteen bushels, the yield of clover has decreased from two and eight-tenths tons per acre to less than one-half ton, while the yield of wheat has decreased only from thirty bushels to twenty-four bushels. As a general average the late yields are only one-third as large as they were fifty years before on the same land. Wheat grown once in four years has been the only crop worth raising on the unfertilized land during the last twenty years, and even the wheat crop has distinctly decreased in yield; although where mineral plant food was applied the yield has increased from thirty bushels, in 18851 to thirty-eight bushels as an average of the last twenty years. In the fallow rotation on the unfertilized land the yield of wheat averaged thirty-four and five-tenths bushels during the first twenty years (1848 to 1867) and twenty-three and five-tenths bushels during the last twenty years.

"On another Rothamsted field the phosphorus actually removed in fifty-five crops from well-fertilized land is two-thirds as much as the total phosphorus now contained in the plowed soil of adjoining untreated land.

"In the early 80's the Pennsylvania Agricultural Experiment Station began a four-year crop rotation, including corn, oats, wheat, and mixed clover and timothy.

"There are five plots in each of four different fields that have received no applications of plant food from the beginning. Thus, every year the crops are carefully harvested and weighed from twenty measured plots of ground that receive no treatment except the rotation of crops. The difference between the average of the first twelve years and the average of the second twelve years should represent the actual change in productive power during a period of twelve years. These averages show that the yield of corn has decreased from forty-one and seven-tenths bushels to twenty-seven and seven-tenths bushels; that the yield of oats has decreased from thirty-six and seven-tenths bushels to twenty-five bushels; that the yield of wheat has decreased only from thirteen and three-tenths bushels to twelve and eight-tenths bushels; and that the yield of hay has decreased from three thousand seventy pounds to two thousand one hundred and eighty pounds.

"As a general average of these four crops the annual value of produce from one acre has decreased from $11.05 to $8.18. Here we have information which is almost if not quite equal in value to that from the Agdell rotation field at Rothamsted. While the Rothamsted experiments cover a period of sixty years, each crop was grown but once in four years; whereas, in the Pennsylvania experiments, there have been four different series of plots, so that in twenty-four years there have been twenty-four crops of corn, twenty-four crops of oats, twenty-four crops of wheat, and twenty-four crops of hay.

"Under this four-year rotation the value of the crops produced has decreased twenty-six per cent. in twelve years. What influence will impress that fact upon the minds of American landowners? A loss amounting to more than one-fourth of the productive power of the land in a rotation with clover seeded every fourth year! This one fact is the mathematical result of four hundred and eighty other facts obtained from twenty different pieces of measured land during a period of twenty-four years.

"As an average of these twenty-four years, the addition of mineral plant food produced increases in crop yields above the unfertilized land as follows:

Corn increased forty-five per cent.
Oats increased thirty-two per cent.
Wheat increased forty-two per cent.
Hay increased seventy-seven per cent.

"As a general average of the four crops for the twenty-four years, the produce where mineral plant food is applied, was forty-nine per cent. above the yields of the unfertilized land, although the same rotation of crops was practiced in both cases."

"Those are some of the absolute facts of science secured from practical application in the adoption and development of definite systems of permanent prosperous agriculture, and they should be made to serve this greatest and most important industry just as the established facts of mathematical and physical science are made to serve in engineering."

"I am glad to know about those long-continued experiments," said Adelaide. "They are of fascinating interest. I have been so sorry for grandma, and for papa and mamma, because of their increasing discouragement over our farm. I do hope we may profit from this fund of accumulated information which has already been secured from long years of investigation. Surely we must endeavor to avoid in America the awful conditions that already exist in the older agricultural countries, where the lands are depleted and the people are brought to greater poverty than even here in Virginia.

"But we have already reached the turn, and you have a mile to walk. How much time have you?"

"Thirty minutes yet," said Percy. "Wait just a moment. Have you read
Lincoln's stories?"

"Many of them, yes."

"Here is the best one he ever told; I have copied it on a card. He told it to a meeting of farmers at the close of an address in which he urged them to study the science of agriculture and to adopt better methods of farming:

"'An Eastern monarch once charged his wise men to invent him a sentence to be ever in view, and which should be true and appropriate in all times and situations. They presented him the words, "And this, too, shall pass away." How much it expresses! How chastening in the hour of pride! How consoling in the depths of affliction! "And this, too, shall pass away." And yet, let us hope, it is not quite true. Let us hope, rather, that by the best cultivation of the physical world beneath and around us, and the best intellectual and moral world within us, we shall secure an individual, social, and political prosperity and happiness, whose course shall be onward and upward, and which, while the earth endures, shall not pass away.'"

"I agree with you that it is his best story," said Adelaide, as Percy finished reading and placed the card in her hand. "Now you must go or I shall insist upon taking you back to the station."

"I shall stand here and time you till you reach the next turn," he replied. "Then you will be in sight of Westover. One! Two! Three! Go!"

CHAPTER XXXIII

THE DIAGNOSIS AND PRESCRIPTION

WINTERBINE, ILLINOIS,

December 4, 1 903

Mr. T. O. Thornton, Blairville, VA.

MY DEAR SIR:—I beg to report that I returned home a few days ago and found my mother well and busy as usual. We have definitely decided that we will not accept your kind offer to sell us a part of your farm, but we appreciate nevertheless the sacrifice, at least from the standpoint of sentiment, which Mrs. Thornton and Miss Russell were willing to make, in order to permit us to secure such a farm as we might want in a splendid situation.

As a matter of fact we are thinking very seriously of purchasing a farm in Southern Illinois. My mother much prefers to remain in Illinois, and for some reasons I have the same preference on her account.

While in Washington I was fortunate enough to find that a soil survey had been completed for your county and also that a partial ultimate analysis had been made of the common loam soil of your farm, such as we sampled. Following are the number of pounds per acre for the surface soil to a depth of six and two-thirds inches,—that is, for two million pounds of soil.

610 pounds of phosphorus

13,200 pounds of potassium

1,200 pounds of magnesium

3,430 pounds of calcium

As compared with a normal fertile soil, your land is very deficient in phosphorus and magnesium, and, as you know, the soil is acid. It is better supplied with potassium than with any other important element.

I would suggest that you make liberal use of magnesian limestone,—at least two tons per acre every four or five years,—and the initial application might better be five or even ten tons per acre if you are ready to make such an investment.

I am sorry that the nitrogen content of the soil was not determined, or at least not published in the bulletin. There can be no doubt, however, that your soil is extremely deficient in organic matter and nitrogen, and you will understand that liberal use should be made of legume crops. The known nitrogen content of legumes and other crops will be a help to you in planning your crop rotation and the disposition of the crops grown.

As to phosphorus, it is safe to say that in the long run fine-ground rock phosphate will prove the best investment; but for a few years it might be best to make some use of acid phosphate in addition to the raw rock, at least until you are ready to begin turning under more organic matter with the phosphate.

There is only one other suggestion: If you wish to make a start toward better crops as soon as possible, you may well use some kainit,—say six hundred pounds per acre every four or five years, preferably applied with the phosphate. In the absence of decaying organic matter, the potassium of the soil becomes available very slowly. The kainit furnishes both potassium and magnesium in soluble form and it also contains sulfur and chlorin. As soon as you can provide plenty of decaying organic matter you will probably discontinue the use of both kainit and acid phosphate. If you sell only grains and animal products, the amount of potassium sold from the farm is very small compared with your supply of that element, which would be sufficient for one hundred bushels of corn per acre for seven hundred years.

I have some doubt if it will be worth the expense involved to have the samples of subsurface and subsoil analyzed at this time; but you might save them for future use if desired.

I shall always appreciate the kindness shown me by being permitted to enjoy your hospitality and to profit from the information you were so able to give me concerning the history and general character of your lands.

My mother asks to have her kind regards extended to you and yours.

Very sincerely yours,

PERCY JOHNSTON.

WESTOVER, January 2, 1904. Percy Johnston, Esq., Winterbine, Ill.

MY DEAR FRIEND:—We were all pleased to receive your letter informing us of your safe journey back to Illinois. I had hoped that you might find a piece of land here in the East which would suit you; but I am not surprised that you and your mother should prefer to remain in Illinois, because of your former associations and your better knowledge of the Western conditions. Northern men who come South often have serious difficulty to manage our negro labor.

I am surprised, however, that you were able to purchase, even in Southern Illinois, such prairie land as you describe for the price of $18 per acre. I supposed $190 an acre for your corn belt farm was a good price, although it is commonly reported to us that Illinois land is selling for $150 to $200 an acre.

Now, in regard to correspondence with Adelaide, let me say that we could have no objection whatever, except that it might be misunderstood, more especially, of course, by Professor Barstow. I do not think I mentioned it to you, but the fact is that the Professor and Adelaide are essentially betrothed. I do not know that the final details are perfected, but doubtless they are, for they have been much together during the Christmas weeks. The Barstows, as you probably know, are still among the most prominent people of North Carolina. Adelaide is young yet and we respect her reticence, but her mother and I have both given our consent and Professor Barstow has every reason to be satisfied with the reception he invariably receives from Adelaide.

I only mention this matter to you that you may understand why misunderstanding might arise in case of such correspondence as you suggest, even though, as Adelaide has explained, she has very naturally become interested temporarily in some of the economic and social questions relating to agriculture, and would unquestionably read your letters concerning these state and national problems with continued interest. I shall hope, however, that she may still have that satisfaction, for I am very deeply interested in all such questions, and I am particularly interested to know more of the details of your southern Illinois farm, including the invoice of the soil, which you say has been taken by your Experiment Station, and especially your definite plans for the improvement of the land. I hope the name you have chosen for your farm is not so appropriate as it would be for some of our old Virginia farms.

I shall also be under renewed obligation to you if I may occasionally submit questions concerning the best plans for the restoration of Westover to its former productiveness. I have decided at least to make another trial with alfalfa next summer, following the valuable suggestions you gave me.

In closing let me renew my assurance of our deep gratitude for the special service you so nobly rendered when fiendish danger threatened my daughter. We shall always regard you as a gentleman of the highest type. Very respectfully yours,

CHARLES WEST.

Percy read this letter hurriedly to the end, and then slowly reread it. His mother noticed that he absent-mindedly replaced the letter in the envelope instead of reading it to her as was his custom. However, he laid the letter by her plate and talked with her about the corn-shelling which was to begin as soon as the corn sheller could be brought from the neighbor's where Percy had been helping to haul the corn from the sheller to elevator at Winterbine. Dinner finished, he hurried out to complete the preparations for the afternoon's work. We have no right to follow him. His mother only saw that he went to the little granary where a few loads of corn were to be stored for future use. Yes, she saw that he closed the door as he entered. Not even his mother could see her son again a child. Women and children weep, not men. The heart strings draw tight and tighter until they tear or snap. The body is racked with the anguish of the mind. The form reels and sinks to the floor. The head bows low. Pent up tears fall like rain.—No, that cannot be. Men do not shed tears. If they are mental cowards and physical brutes they pass from hence by a short and easy route and leave the burdens of life to their wives and mothers and disgraced families. If they are Christian men they seek the only source of help.

Mrs. Johnston watched and waited—it seemed an hour, but was only a quarter of that time till the granary door opened and she saw Percy pass to the barn with a step which satisfied her mother's eye.

She drew out the letter, and from a life habit of making sure, pressed the envelope to see that it contained nothing more. She noted a slip of crumpled paper and drew it out. Upon it was written in a penciled scrawl:

"*Her grandma has not consented.*"

She read the letter, stood for a moment as in meditation, then replaced the slip and letter in the envelope, and laid it on Percy's desk. The letter was plainly a man's handwriting. The envelope was addressed in a bold hand that was clearly not Mr. West's writing.

CHAPTER XXXIV

PLANNING FOR LIFE

HEART-OF-EGYPT, ILLINOIS, June 16, 1904.

Mr. Charles West,

Blue Mound, Va.

MY DEAR SIR:—I have delayed writing to you in regard to the plans for Poorland Farm, until I could feel that we are able at least to make an outline of tentative nature. The labor problem of a farm of three hundred and twenty acres is of course very different from that on forty acres, and we are not yet fully decided regarding our crop rotation and the disposition of the crops produced (or hoped for). I realize that to rebuild in my life what another has torn down during his life is a task the end of which can hardly be even dimly foreshadowed. Some friends are already beginning to ask me what results I am getting, and they apparently feel that we must succeed or fail with a trial of a full season. I have said to them that I have no objection whatever to discussing our plans at any time, so far as we are yet able to make plans, but that I shall not be ready to discuss results with anyone until we begin to secure crop yields in the third rotation. This means that I am not expecting the benefits of a six-year rotation of crops before the rotation has been actually practiced. You will understand of course that, if all your land had been cropped with little or no change, for all its history, you would require six or eight years' time before you would be able to grow a crop of corn on land that had been pastured for six or eight years;

but some people seem to take it for granted that one can adopt a six-year rotation and enjoy the full benefits of it the first season.

I remember that you were surprised that I could buy a level upland farm even in this part of Illinois for $18 an acre; but you will probably be more surprised to learn that this farm had not paid the previous owners two per cent. interest on $18 an acre as an average of the last five years. In fact, sixty acres of it had grown no crops for the last five years. It was largely managed by tenants on the basis of share rent, and because of this I have been able to secure the records of several years.

I at least had some satisfaction in purchasing this farm, for the real estate men were left without a single "talking point." I insisted that I wanted the poorest prairie farm in "Egypt," and whenever they began to tell me that the soil on a certain farm was really above the average, or that the land had been well cared for until recently, or that it had been fertilized a good deal, etc., I at once informed them that any advantage of that sort completely disqualified any farm for me; and that they need not talk to me about any farms except those that represented the poorest and most abused in Southern Illinois.

I may say, however, that $20 an acre is about the average price of the average land. I had an option on a three hundred and sixty acre farm cornering the corporation limits of the County Seat for $30 an acre, and all agreed that the farm was above the average in quality.

Heart-of-Egypt is a small station on the double track of the Chicago-New Orleans line of the Illinois Central, and there are three other railroads passing through our County Seat. Poorland Farm is less than two miles from Heart-of-Egypt and only five miles from the County Seat, with level roads to both.

As to the soil, I may say that in some respects it is poorer than yours, but in others not so poor. The amount of plant food contained in six and two-thirds inches of the surface soil of an acre, representing two million pounds of soil, are as follows:

2,880 pounds of nitrogen

840 pounds of phosphorus

24,940 pounds of potassium

6,740 pounds of magnesium

14,660 pounds of calcium

By referring to the invoice of your most common land, you will see that Westover is richer in phosphorus, in magnesium, and in calcium, than Poorland Farm. But, while your soil contains a half more of that rare element phosphorus, ours contains a half more of the abundant element potassium. In the supply of nitrogen we have a distinct advantage, because our soil contains nearly three times as much as your most common cultivated land, and even twice as much as your level upland soil, which you consider too poor for farming, but in which phosphorus and not nitrogen must be the first limiting element, the same as with ours.

The fact is that the nitrogen problem in the East was one of the reasons why we have chosen to locate in Southern Illinois. I am confident that the level lands I saw about Blairville and over in Maryland are more deficient in organic matter and nitrogen than your uncultivated level upland, and probably even more deficient than your common gently sloping cultivated lands, because of your long rotation with much opportunity for nitrogen fixation by such legumes will grow in your meadows and pastures, including the red clover which you regularly sow, the white clover, which is very persistent, and the Japan clover, which it seems to me has really benefited you more than the others.

To me a difference in nitrogen content of two thousand pounds per acre signifies a good deal. It plainly signifies a hundred years' of "working the soil for all that's in it," beyond what has yet been done to our "Egypt." The cost of two thousand pounds of nitrogen in sodium nitrate would be at least $300 and even that would not include the organic matter, which has value for its own sake because of the power of its decomposition products to liberate the mineral elements from the soil, as witness the most common upland soils of St. Mary county, Maryland, with a phosphorus content reduced to one hundred and sixty pounds per acre in two million pounds of the ignited soil. The ten-inch plows of Maryland, the twelve-inch of Southern Illinois, the fourteen-inch of the corn belt, and the sixteen-inch of the newer regions of the Northwest, signify something as to the influence of organic matter upon the horsepower required in tillage; and the organic matter also has a value because it increases the power of the soil to absorb and retain moisture and to resist surface washing and "running together" to form the hard surface crust.

To think of applying two thousand pounds of nitrogen by plowing under two hundred tons of manure or forty tons of clover per acre at least requires a "big think," as my Swede man would say.

Of course, with our western life and cosmopolitan population, where "a man's a man for a' that," mother feels that it would not be easy for us to fit into your somewhat distinctly stratified society. We would not be "colored" if we could, and perhaps we could not be aristocratic if we would; and the opportunity to become, or, perhaps I should say, to remain, "poor white trash," though wide open, is not very alluring. I realize, of course, that there are some whole-souled people like the West's and Thornton's, but I also found some of the tribe of Jones, and I have much doubt as to the social standing of one who would feel obliged to demonstrate that he could spread more manure in a day than his hired nigger.

My Swede and I are like brothers; we clean stables together and talk politics, science, and agriculture. In fact he is as much interested as I am in the building up of Poorland Farm, and has already contributed some very practical suggestions. I pay him moderate wages and a small percentage of the farm receipts after deducting certain expenses which he can help to keep as low as possible, such as for labor, repairs, and purchase of feed and new tools, but without deducting the taxes or interest on investment or the cost of any permanent improvements, such as the expense for limestone, phosphate, new fences and buildings, and breeding stock.

Referring again to the invoice of the soil, I may say that the percentage of the mineral plant foods increases with depth, the same as in your soil, but not to such an extent, and with one exception. The phosphorus content of our surface soil is greater than that of the subsurface, but below the subsurface the phosphorus again increases. This is probably due to the fact that the prairie grasses that grew here for centuries extracted some phosphorus from the subsurface in which their roots fed to some extent, and left it in the organic residues which accumulated in the surface soil.

Aside from the difference in organic matter, the physical character of our soil is distinctly inferior to the loam soils about Blairville and Leonardtown. We have a very satisfactory silt loam surface, but the structure of our subsoil is quite objectionable. It is a tight clay through which water passes very slowly, so slowly that the practicability of using tile-drainage is still questioned by the State University, although the experiments which the University soil investigators have already started in several counties here in "Egypt" will ultimately furnish us positive knowledge along this line.

As for me, I purpose making no experiments, whatever. I do not see how I or any other farmer can afford to put our limited funds into experiments, especially when we often lack the facilities for taking the exact and complete data that are needed. It takes time and labor and some equipment to make accurate measurements, to weigh every pound of fertilizer applied and every crop carefully harvested from measured and carefully seeded areas, especially selected because of their uniform and representative character. I think this is public business and it is best done by the State for the benefit of all.

I have heard narrow politicians call it class legislation to appropriate funds for such agricultural investigations, but the fact is that to investigate the soil and to insure an abundant use of limestone, phosphate, or other necessary materials required for the improvement and permanent maintenance of the fertility of the soil is legislation for all the people, both now and hereafter. Would that our Statesmen would think as much of maintaining this most important national resource, as they do of maintaining our national honor by means of battleships and an army and navy supported at an expense of three hundred million dollars a year, sufficient to furnish ten tons of limestone to every acre of Virginia land, an amount twenty times the Nation's appropriation for agriculture; and even this is largely used in getting new lands ready for the bleeding process, instead of reviving those that have been practically bled to death.

As for me, I shall simply take the results which prove profitable on the accurately conducted experiment fields of the University of

Illinois, one of which is located only seven miles from Poorland Farm, and on the same type of soil, I shall try to profit by that positive information, and await the accumulation of conclusive data relating to tile-drainage and other possible improvements of uncertain practicability for "Egypt."

Say, but our soil is acid! The University soil survey men say that the acidity is positive in the surface, comparative in the subsurface, and superlative in the subsoil. Two of them insisted that the subsoil has an acid taste. The analysis of a set of soil samples collected near Heart-of-Egypt shows that to neutralize the acidity of the surface soil will require seven hundred and eighty pounds of limestone per acre, while three tons are required for the first twenty inches, and sixteen tons for the next twenty inches. The tight clay stratum reaches from about twenty to thirty-six inches. Above this is a flour-like gray layer varying in thickness from an inch to ten inches, but below the tight clay the subsoil seems to be more porous, and I am hoping that we may lay tile just below the tight clay and then puncture that clay stratum with red clover roots and thus improve the physical condition of the soil. I asked Mr. Secor, a friend who operates a coal mine,—and farms for recreation,—if he thought alfalfa could be raised on this type of soil. He replied: "That depends on what kind of a gimlet it has on its tap root."

Some of the farmers down here tell me confidentially that "hardpan" has been found on their neighbors' farms, but I have not talked with any one who has any on his own farm. I am very glad the University has settled the matter very much to the comfort of us "Egyptians," by reporting that no true "hardpan" exists in Illinois, although there are extensive areas underlain with tight clay, "of whom, as it were, we are which."

I am glad that the nitrogen-fixing and nitrifying bacteria do business chicfly in the surface soil, because we are not prepared to correct the acidity to any very great depth.

The present plan is to practice a six-year rotation on six forty-acre fields, as follows:

First year—Corn (and legume catch crop).

Second year—Part oats or barley, part cowpeas or soy beans.

Third year—Wheat.

Fourth year—Clover, or clover and timothy.

Fifth year—Wheat, or clover and timothy.

Sixth year—Clover, or clover and timothy.

This plan may be a grain system where wheat is grown the fifth year, only clover seed being harvested the fourth and sixth years, or it may be changed to a live-stock system by having clover and timothy for pasture and meadow the last three years, which may be best for a time, perhaps, if we find it too hard to care for eighty acres of wheat on poorly drained land.

In somewhat greater detail the system may be developed we hope about as follows:

First year: Corn, with mixed legumes, seeded at the time of the last cultivation, on perhaps one-half of the field. These legumes may include some cowpeas and soy beans and some sweet clover, but that is not yet fully decided upon.

Second year: Oats (part barley, perhaps) on twenty acres, cowpeas on ten acres, and soy beans on ten acres. The peas and beans are to be seeded on the twenty acres where the catch crop of legumes is to be plowed under as late in the spring as practicable.

Third year: Wheat with alsike on twenty acres and red clover on the other twenty, seeded in the early spring. If necessary to prevent the clover or weeds from seeding, the field will be clipped about the last of August.

Fourth year: Harvest the red clover for hay and the alsike for seed, and apply limestone after plowing early for wheat.

Fifth year: Wheat, with alsike and red clover seeded and clipped as before.

Sixth year: Pasture in early summer, then clip if necessary to secure uniformity, and later harvest the red clover for seed. Manure may be applied to any part of this field from the time of wheat harvest the previous year until the close of the pasture period. Then it may be applied to the alsike only until the red clover seed crop is removed, and then again to any part of the field, which may also be used for fall pasture. To this field the threshed clover straw and all other straw not needed for feed and bedding will be applied. The application of raw phosphate will be made to this field, and all of this material plowed under for corn.

The second six years is to be a repetition of the first, except that the alsike and red clover will be interchanged, so as to avoid the development of clover sickness if possible; and to keep the soil uniform we may interchange the oats with the peas and beans.

This system provides for the following crops each year:

40 acres of corn;

20 acres of oats;

10 acres of cowpeas for hay

10 acres of soy beans for seed

80 acres of wheat

20 acres of red clover for hay

20 acres of alsike for seed

20 acres of red clover for seed

20 acres of alsike for pasture, except from June to August.

We also have some permanent pasture which we may use at any time that may seem best. If necessary we may cut all the clover for hay the fourth year, and we may pasture all summer the sixth year. We can pasture the corn stalks during the fall and winter when the ground is in suitable condition.

We plan to raise our own horses and perhaps some to sell. In addition we may raise a few dairy cows for market, but will do little dairying ourselves.

We expect to sell wheat and some corn, and if successful we shall sell some soy beans, alsike seed, and red clover seed.

How soon we shall be able to get this system fully under way I shall not try to predict; but we shall work toward this end unless we think we have good reason to modify the plan.

I hope to make the initial application of limestone five tons per acre, but after the first six years this will be reduced to two or three tons. I also plan to apply at least one ton per acre of fine-ground raw phosphate every six years until the phosphorus content of the plowed soil approaches two thousand pounds per acre, after which the applications will probably be reduced to about one-half ton per acre each rotation.

There are three things that mother and I are fully decided upon:

First, that we shall use ground limestone in sufficient amounts to make the soil a suitable home for clover.

Second, that we shall apply fine-ground rock phosphate in such amounts as to positively enrich our soil in that very deficient element.

Third, that we shall reserve a three-rod strip across every forty-acre field as an untreated check strip to which neither limestone nor phosphate shall ever be applied, and that we shall reserve another three-rod strip to which limestone is applied without phosphate, while the remaining thirty-seven acres are to receive both limestone and phosphate.

Thus we shall always have the satisfaction of seeing whatever clearly apparent effects are produced by this fundamental treatment, even though we may not be able to bother with harvesting these check strips separate from the rest of the field.

We have based our decision regarding the use of ground limestone very largely upon the long-continued work of the Pennsylvania Agricultural Experiment Station as to the comparative effects of ground limestone and burned lime, which is supported, to be sure, by all comparative tests so far as our Illinois soil investigators have been able to learn.

The practicability and economy of using the fineground natural phosphate has been even more conclusively established, as you already know, by the concordant results of half a dozen state experiment stations. There are only two objections to the use of the raw phosphate. One of these is the short-sighted plan or policy of the average farmer, and the other is the combined influence of about four-hundred fertilizer manufacturers who prefer to sell, quite naturally, perhaps, two tons of acid phosphate for $30, or four tons of so-called "complete" fertilizer for $70 to $90, rather than to see the farmer buy direct from the phosphate mine one ton of fine-ground raw rock phosphate in which he receives the same amount of phosphorus, at a cost of $7 to $9.

Until we can provide a greater abundance of decaying organic matter we may make some temporary use of kainit, in case the experiments conducted by the state show that it is profitable to do so.

In a laboratory experiment, made at college it was shown that when raw phosphate was shaken with water and then filtered, the filtrate contained practically no dissolved phosphorus; but, if a dilute solution of such salts as exist in kainit was used in place of pure water, then the filtrate would contain very appreciable amounts of phosphorus.

In addition to this benefit, the kainit will furnish some readily available potassium, magnesium, and sulfur; and, by purchasing kainit in carload lots, the potassium will cost us less than it would in the form of the more expensive potassium chlorid or potassium sulfate purchased in ton lots. Of course we do not need this in order to add to our total stock of potassium, but more especially I think to assist in liberating phosphorus from the raw phosphate which is naturally contained in the soil and which we shall also apply to the soil, unless the Government permits the fertilizer trusts to get such complete control of our great natural phosphate deposits that they make it impossible for farmers to secure the fine-ground rock at a reasonable cost, which ought not, I would say, to be more than one hundred per cent. net profit above the expense of mining, grinding, and transportation. We may feel safe upon the matter of transportation rates, for the railroads are operated by men of large enough vision to see that the positive and permanent maintenance of the fertility of the soil is the key to their own continued prosperity, and some of them are already beginning to understand that the supply of phosphorus is the master key to the whole industrial structure of America; for, with a failing supply of phosphorus, neither agriculture nor any dependent industry can permanently prosper in this great country.

If we retain the straw on the farm and sell only the grain, the supply of potassium in the surface soil of Poorland Farm is sufficient to meet the needs of a fifty bushel crop of wheat per acre every year for nineteen hundred and twenty years, or longer than the time that has passed since the Master walked among men on the earth; whereas, the total phosphorus content of the same soil is sufficient for only seventy such crops, or for as long as the full life of one man. Keep in mind that Poorland Farm is near Heart-of-Egypt, and that this is the common soil of our "Egyptian Empire," which contains more cultivable land than all New England, has the climate of Virginia, and a network of railroads scarcely equalled in any other section of this country, and in addition it is more than half surrounded by great navigable rivers.

On Poorland Farm there are seven forty-acre fields which are at least as nearly level as they ought to be to permit good surface drainage, and there is no need that a single hill of corn should be omitted on any one of these seven fields; and I am confident that with an adequate supply of raw phosphate rock and magnesian limestone and a liberal use of legume crops this land can be made to pay interest on $300 an acre.

Why not? At Rothamsted, England, they have averaged thirty-eight and four-tenths bushels of wheat per acre during the last twenty years in an experiment extending over sixty years, and they have done this without a forkful of manure or a pound of purchased nitrogen. Why not? The wheat alone from eighty acres of land, if it yielded forty bushels per acre and sold at $1 a bushel, would pay nearly five per cent. interest on $300 an acre for the entire two hundred and forty acres used in my suggested rotation.

Aye, but there is one other very essential requirement: To wit, a world of work.

Hoping to hear from you, and especially about your alfalfa, I am,

Very sincerely yours,

PERCY JOHNSTON.

CHAPTER XXXV

SEALED LIPS

No one realized more than Percy Johnston that toleration of life itself was possible to him only because of the world of work that he found always at hand in connection with his abiding faith and interest in the upbuilding of Poorland Farm. He had accepted Adelaide's sweet smile and lack of apparent disapproval with confidence that he might at least have an opportunity to try to win her love. As he was permitted at the parting to look for more than an instant into those alluring eyes, he felt so sure that they expressed something more than friendship or gratitude for him. He had felt the more confidence because he thought he knew that she would not permit him to humiliate himself by asking and failing to receive from her father permission to write to her, when she could easily in her own womanly way have discouraged such a thought at once. Had she not insisted upon driving slowly back to the turn in the road, and did he not feel the absence of a previous reserve?

Oh, misleading imagination. The will is truly the father of thought and faith. Percy knew as he parted from Adelaide that he had left with her the love of heart and mind of one whose life had developed in him the character which does nothing by halves. His love had multiplied with the distance as he journeyed westward, with a great new pleasure which life seemed to hold before him and with a pardonable confidence in its achievement.

He had written Mr. West a week after his return in a way which would not fail of understanding if his hopes were justified. The belated reply which reached him after holidays was accepted as final. His pride was humiliated and the sweetest dream of his life abruptly ended. He felt the more helpless and the more deeply wounded because of Mr. West's reference to his special service in the protection he had once rendered to Adelaide. It continually reminded him that, as the highest type of gentleman, he should do nothing that could be construed as an endeavor to take advantage of the consideration to which that act might seem to entitle him. Bound and buried in the deepest dungeon, waiting only for the announcement from his of the day of his execution. This was his mental attitude as the months passed and he began to receive an occasional letter from Mr. West, in each of which he looked for the news of Adelaide's marriage.

In Mrs. Johnston a feeling of hatred had developed for Adelaide. She was certain that she had marred the happiness of her son. The heartlessness of a flirt who could trifle with the affection of one who had a right to assume in her an honor equal to his own deserved only to be hated with even righteous hatred. She saw the scrawled note which she knew Percy had not seen, but what did it signify? An eccentric old lady's penchant for match making? Perhaps she was even more guilty than the girl in attempting to lead Percy to see in Adelaide more than he ought. She might even take an old flirt's delight in the mere number of conquests made by her granddaughter. Or was the scrawled note slipped into the envelope by a prank-playing fourteen-year-old brother? In any case was it wise that Percy should see the note? She could probably do nothing better than to leave it with the letter. Even if the girl were worthy, Percy could never hope to win one of her class, whose pride of ancestry is their bread of life. It might not have been quite so, perhaps, if Percy had only selected some more respected profession. Why should not he have become a college professor?

CHAPTER XXXVI

HARD TIMES

WHEN Percy and his mother reached Poorland Farm in March they found a small frame house needing only shingles, paint, and paper to make it a fairly comfortable home, until they should be able to add such conveniences as Percy knew could be installed in the country as well as in the city. From the sale of corn and some other produce they were able to add to the residue of $1,840, which represented the difference between the cost of three hundred and twenty acres in Egypt and the selling price of forty acres in the corn belt. An even $3,000 was left in the savings bank at Winterbine.

"If we can live," said Percy, "just as the other 'Egyptians' must live, and save our $3,000 for limestone and phosphate, I believe we shall win out. Through the efforts of the Agricultural College and the Governor of the State the convicts in the Southern Illinois Penitentiary have been put to quarrying stone, and large crushers and grinders have been installed, and the State Board of Prison Industries is already beginning to ship ground limestone direct to farmers at sixty cents a ton in bulk in box cars. The entire Illinois Freight Association gave an audience to the Warden of the Penitentiary and representatives from the Agricultural College and a uniform freight rate has been granted of one-half cent per ton per mile. This will enable us to secure ground limestone delivered at Heart-of-Egypt for $1.22-1/2 per ton.

"Now, to apply five tons per acre on two hundred and forty acres will require one thousand two hundred tons and that will cost us

$1,570 in cash, less perhaps the $70, which we save on roads and the untreated check strips which I want to leave. To apply one ton of phosphate per acre to the same six fields will cost about $1,600. Of course, I shall not begin to apply phosphate until after I have applied the limestone and get some clover or manure to mix with the phosphate when I plow it under; and I hope with the help of the limestone we shall get some clover and some increase in the other crops. In any case the $3,000 and interest we will get for what we can leave in the bank during the six or eight years it will take to get the rotation and treatment under way will pay for the initial cost of the first applications of both limestone and phosphate; and we shall hope that by that time the farm will bring us something more than a living."

The carload of effects shipped from Winterbine to Heart-of-Egypt included two horses, a cow, a few breeding hogs, and some chickens; also a supply of corn and oats sufficient for the summer's feed grain.

After the expenses of shipping were paid, less than $350 were deposited in the bank at the County Seat. Of this $250 were used for the purchase of another team. Hay was bought from a neighbor and some old hay that had been discarded by the balers, who had purchased, baled, and sold the previous hay crop from Poorland Farm, Percy gathered up and saved for bedding.

He plowed forty acres of the land that had not been cropped for five years, and, after some serious delays on account of wet weather, planted the field in corn, using the Champion White Pearl variety, be cause the Experiment Station had found it to be one of the best varieties for poor land.

"I wouldn't plant that corn if you would give me the seed," a neighbor had said to him. "See how big the cob is; and the tip is not well filled out, and there is too much space between the rows. I tell you there's too much cob in it for me. I want to raise corn and not corn cob."

"It certainly is not a good show ear," said Percy, "but what I want most is bushels of shelled corn per acre. Perhaps these big kernels will help to give the young plant a good start, and perhaps the piece of cob extending from the tip will make room for more kernels if the soil can be built up so as to furnish the plant food to make them. The cob is large but it is covered with grains all the way around; and, if those kernels of corn were putty, we could mash them down a little and have less space between the rows, but it would make no more corn on the ear. However, my chief reason for planting the Champion White Pearl is that this variety has produced more shelled corn per acre than any other in the University experiments on the gray prairie soil of 'Egypt.'"

There were only sixteen acres of corn grown on the entire farm in 1903 and this yielded thirteen bushels per acre, as Percy learned from the share of the crop received by the previous landowner.

In 1904 the Champion White Pearl yielded twenty bushels per acre, as nearly as could be determined by weighing the corn from a few shocks on a small truck scale Percy had brought from the north. He numbered his six forty-acre fields from one to six. Forty No. 7 was occupied by twelve acres of apple orchard, eight acres of pasture, and twenty acres of old meadow. By getting eighty rods of fencing it was possible to include twenty-eight acres in the pasture, although one hundred and ninety-two rods of fencing had been required to surround the eight-acre pasture. The remainder of the farm was in patches, including about fifteen acres on one corner crossed by a little valley and covered with trees, a tract which Percy and his mother treasured above any of the forty-acre fields. While the week was always filled with work, there were many hours of real pleasure found in the wood's pasture on the Sunday afternoons.

Forty No. I was left to "lie out," and No. 2 raised only twelve acres of cowpeas. No. 3 was plowed during the summer and seeded to timothy in the early fall. No. 4 was in corn and Nos. 5 and 6 were left in meadow, two patches of nine and sixteen acres previously in cowpeas and corn having been seeded to timothy in order, as Percy said, to "square out" the forty-acre fields. About fifty acres of land were cut over for about sixteen tons of hay. The corn was all put in shock, and the fodder as well as the grain used for feed, the refuse from the fodder and poor hay serving as bedding. About three tons of cowpea hay of excellent quality were secured from the twelve acres, and fifty barrels of apples were put in storage.

Another cow and eight calves were bought, and during the winter, some butter, two small bunches of the last spring's pigs, and the apple crop were sold. A few eggs had been sold almost every week since the previous March.

In 1905 No. 1 was rented for corn on shares and produced about six hundred bushels of which Percy received one-third. No. 2 yielded four hundred and eighty-four bushels of oats. No. 3 produced fourteen tons of poor hay. No. 4 was "rested" and prepared for wheat, ground limestone having been applied. No. 5 was fall-plowed from old meadow and well prepared and planted to corn in good time; but, after the second cultivation, heavy rains set in and continued until the corn was seriously damaged on the flat areas of the field, the more so as he had not fully understood the importance of keeping furrows open with outlets at the head-lands through which the excess surface water could pass off quickly under such weather conditions. Patches of the field aggregating at least five acres were so poorly surface drained that the corn was "drowned out," and fifteen acres more were so wet as to greatly injure the crop. However, on the better drained parts of the field where the corn was given further cultivation the yield was good and about 1,000 bushels of sound corn were gathered from the forty acres.

A mixture of timothy, redtop and weeds was cut for hay on No. 6, the yield being better than half a ton per acre.

The apples were a fair crop, and the total sales from that crop amounted to $750, but about half of this had been expended for trimming and spraying the trees, a spraying outfit, barrels, picking, packing, freight and cold storage. A good bunch of hogs were sold.

Another year passed. Oats were grown on No. 1 and on part of No. 2, yielding eleven bushels per acre.

No. 3 yielded one-third of a ton of hay per acre.

Wheat was grown on No. 4, and clover, the first the land had known in many years, if ever, was seeded in the spring,—twenty acres of red clover and twenty of alsike.

The fifty-four acres of wheat, including fourteen acres on No. 2, yielded seven and one-half bushels per acre. Soy beans were planted on No. 5, but wet weather seriously interferred and only part of the field was cut for hay. Limestone was applied, but heavy continued rains prevented the seeding of wheat.

No. 6 produced about twenty-seven bushels per acre of corn.

Two lots of hogs were sold for about $800, and some young steers increased the receipts by nearly $100.

Mrs. Johnston continued to buy the groceries with eggs and butter; but it was necessary to buy some hay, and the labor bill was heavy.

No. 5 joined the twenty-eight acre pasture and on two other sides it joined neighbors' farms where line fences were up, and on the other side lay No. 4.

Percy was trying to get ready to pasture the clover on No. 4, and a mile of new fencing was required. The materials were bought and the fence built, and when finished it also completed the fencing required to enclose No. 5. The twenty-eight acre pasture was inadequate for sixteen head of cattle and the young stock was kept in a hired pasture. Unless he could produce more feed, Percy saw that the farm would soon be overstocked, for some colts were growing and eight cows were now giving milk.

His hope was in the clover, but as the fall came on the red clover was found to have failed almost completely, and the alsike was one-half a stand. As the red clover had been seeded on the unlimed strip there was no way of knowing whether the limestone had even benefited the alsike. The neighbors had "seen just as good clover without putting on any of that stuff."

There were no apples, but the spraying had cost as much as ever, and some team work had been hired.

Three years of the hardest work; limestone on two forties, but only twenty acres of poor clover on one and no wheat seeded on the other. The neighbors "knew the clover would winter kill." The bills for pasturing amounted to as much as the butter had brought; for the twenty-eight-acre pasture had been very poor. The feed for the cows for winter consisted of corn fodder, straw and poor hay, and not enough of that.

They had to do it—draw $150 from the Winterbine reserve, besides what had been used for limestone. Part of it must go for clover seed, for clover must be seeded before it could be grown. The small barn must also be enlarged, but with the least possible expense.

It was February. Wet snow, water, and almost bottomless mud covered the earth. With four horses on the wagon, Percy had worked nearly all day bringing in two "jags" of poor hay from the stack in the field. It was all the little mow would hold.

He had finished the chores late and came in with the milk.

"Put on some dry clothes and your new shoes," said his mother, "while I strain the milk and take up the supper. There is a letter on the table. I hardly see how the mail man gets along through these roads. They must be worse than George Rogers Clark found on his trip from Kaskaskia to Vincennes. They say his route passed across only a few miles from the present site of Heart-of-Egypt. I suppose the letter is from Mr. West."

Percy finished washing his hands, and opened the letter. Two cards fell to the table as he drew the letter from the envelope.

He picked up one of the cards, and read it aloud to his mother:

_Mr. and Mrs. Strongworth Barstow

__At home after March I, 1907

1422 College Avenue

Raleigh, N. C._

"*With Grandma's Compliments,*" was penciled across the top of the card. Percy glanced at the other card and read the plain lines:

Announce the marriage of their daughter

Did his eyes blurr? He laid the one card over the other, scanned Mr. West's letter hurriedly, replaced it with the cards in the envelope, and laid the letter at his mother's plate.

Percy replaced his rubber boots with shoes, and his wet, heavy coat with a dry one.

"You remember the letter I had from the College?" he asked, as he took his seat at the table.

"Yes, I remember," she replied, "but the Institute was to begin to-day."

"I know," said Percy, "but Hoard and Terry both speak to-morrow,—Terry in the morning and the Governor in the afternoon, and they are the men the Professor especially wanted me to hear, if I could. I think I'll 'phone to Bronson's and ask Roscoe to come over and do the chores to-morrow noon. I can get back by nine to-morrow night."

"But, Dear, how in the world can you get to Olney to hear Mr. Terry speak to-morrow morning?"

"There is a train east about eight o'clock," he replied. "Of course the roads are too awful to think of driving to the station, especially since the mares ought not to be used much. I put four on the wagon to-day and tried to be as careful as possible but it does not seem right to use them. I can manage all right. I will get up a little early in the morning and get things in shape so I can leave here by daylight and I am sure I can make the B. & O. station by eight o'clock easily. I will wear my rubber boots and carry my shoes in a bundle. I can change at the depot and put my boots on again when I get back there at seven at night. If it clears up, I will have the moon to help coming home."

But, Percy, you do not mean to walk five miles and back through all this mud and water?"

"I wish you would not worry, Mother. There is grass along the sides most of the way, and I am used to the mud and water. I will spy out the best track as I go in the morning and just follow my own trail coming back."

"Then it is time we were asleep," replied the mother.

CHAPTER XXXVII

HARDER TIMES

THE State Superintendent of Farmers' Institutes called the meeting to order soon after Percy entered the Opera House at Olney about ten o'clock the next morning.

"Divine blessing will be invoked by Doctor T. E. Sisson, pastor of the First Methodist Church of Olney:"

"Oh, Thou, whose presence bright all space doth occupy and all motion guide, all life impart, we come this morning in the capacity of this Farmers' Institute to thank thee for Thy mercies and for Thy blessings, and to invoke Thy presence and Thy continued favor. As Thou with Thy presence hast surrounded all forms of creation and all stages of being with the providences of welfare and development and grace, so we pray, our Father, for guidance through the sessions of this institute, for the providences of Thy love and Thy wisdom divine as it reveals itself in the open field, in the orchard, in the garden, bringing forth those things which replenish the earth with food and fill the mouths of our hungry ones with bread.

"We thank Thee for this larger knowledge which has come to the minds of men, because they have been learning to study Thy works and to walk closer to Thee. Wilt Thou, Heavenly Father, continue to enlighten this body of men and women that are represented in this great field of the world's busy hive so that the starving millions of the world, now in our cities rioting for bread, and in the vast nations where they are crying for food, may be fed. We pray Thee, reveal such improvement of knowledge to these who are willing to get close to Thee to learn Thy secrets and know Thy wisdom, as that unto all shall be given plenty, for replenishing our physical needs. And help us to know, our Father, as we learn Thy will and seek to do Thy will and live in the higher courts of knowledge and wider circles of thought, so shall God reveal himself unto us.

"Our Father, we thank Thee for all the developments and great sources of utility that come through the means of this institute in the development of the resources of this country, this great State and adjoining states through the length and breadth of this favored nation. We pray, Heavenly Father, while studying all these replenishments and seeking to defend them from the inroads of evil, of the rust and the mildew and the worm, we pray also for the beautiful homes, for the souls of the children given to our homes, that we may study their mental and spiritual being in such a way as shall keep all harm and evil and wrong from this life of ours, and so to work in the field of Thy providences, revealed in hand and mind and heart and relationships, of school and church and state and farm, and all the activities of this life's great work, as that good shall be our inheritance.

"We pray Thee, Heavenly Father, to be with the officers of this institute. Give Thy strength, Thy presence, and Thy discernment to these who participate in the work, the membership and onlookers, and those who come to learn. We pray Thee, give us the revelation of Thy wisdom to replenish and build up every human family, and to Thee all praise shall be given to-day for this blessing and for Thy continued favor; and not only to-day but to-morrow and the day after and through all eternity the praise shall be Thine, in the name of Him who came into this world to give us the life of the knowledge of God. Amen."

"It may be," said the Chairman, "that a State Farmers' Institute sometimes exercises a little arbitrary power in selecting subjects we want to speak of. I think county institutes might adopt the same plan to advantage, and assign the topic they wish discussed.

"The topic assigned our speaker to-day is 'What I did and how I did it.' It may sound egotistical, but I want to relieve the speaker of that imputation, because the subject was selected by the Institute.

"Allow me to present Mr. Terry, who needs no introduction to an audience of American farmers:"

Mr. Terry began to speak:

"Thirty-six years ago last fall," he said, "my wife and I bought and moved onto the farm where we now reside. We went on there in debt $3,700, on which we had to pay seven per cent. interest. I had one horse, an old one, and it had the heaves, a one-horse harness, and a one-horse wagon, three tillage implements, and nine cows that were paid for; and a wife and two babies, but no money. Now that was the condition in which we started on this farm, thirty-six years ago, in debt heavily, and no money; but that is not the worst of it. If it had been as good soil as you have in some parts of this State, we should have been all right. How about the soil? For sixty years farmers had been running it down until it could scarcely produce anything. We had a tenant on the place one year, before we could arrange to move on, after we got it. They got eight bushels of wheat per acre, and he said to me, 'That is a pretty good yield, don't you think, for this old farm?' Oh, friends, I didn't think so;—never ought to have bought this farm;—didn't know any better,—born and brought up in town, my father a minister, and I thought a farm was a farm. But I learned some things after awhile. That tenant mowed over probably forty acres of land. (We originally bought one hundred and twenty-five.) He put the hay in the barn. It measured twelve tons. Half of that was weeds. Most of the hay he cut down in a swale. There wasn't anything worth considering on the upland. That was the condition of the land.

"How about the buildings? The house had been used about sixty years, an old story-and-a-half house. Dilapidated, oh, my! Every time the rain came, we had to take every pan upstairs and set it to catch the water. We did not have any money to put on more shingles. It was out of the question, we couldn't do it. How about the dooryard? It was a cow yard. They used it for a milking yard, for years and years. You can imagine how it looked. The barn was in such condition that cattle were just as well off outdoors as in. The roof leaked terribly. The tenants had burned up the doors and any boards they could take off easily. They were too lazy to take off any that came off hard. They burned all the fences in reach.

"Now friends, that was the farm we moved onto and the condition it was in. Some of you will know we saw some pretty hard times for a while. Time and again I was obliged to take my team, after we got two horses (the second I borrowed of a relative, it was the only way I could get one), and go to town to do some little job hauling to get some money to get something to eat. That is the way we started farming. I remember, after three or four years, meeting Dr. W. I. Chamberlain. Some of you know him. He said: 'Terry, if you should get a new hat, there wouldn't anybody know you. Your clothes wear like the children of Israel's.' They had to wear. No one knew how hard up we were. It was not best to let them know. That money was borrowed of a friend in Detroit, secured on a life insurance policy. We did not let anybody know how hard up we really were. My wife rode to town (to church when she went), in the same wagon we hauled out manure in, for a time. Time and again she had been to town when she said she could not do without something any longer and came back without it. Credit was good. We could have bought it. We didn't dare to.

"Now, friends, a dozen years from the time we started on that farm, under these circumstances, we were getting from one hundred and fifty to two hundred and fifty bushels of merchantable potatoes per acre right along—not a single year, but on the average—varying, of course, somewhat with the season. We were getting from four to five tons of clover hay in a season, from two cuttings, of course, per acre. We were getting from thirty-three to thirty-eight bushels of wheat per acre, not one year, but for five years we averaged thirty-five bushels per acre, and right on that same farm. No fertility had been brought on to it, practically, from the outside. A man without any money, in debt for the land $3,700, was able to do this. Now, how did he do it? That is the question I have been asked to talk upon. I have told you briefly something like what we have accomplished. I might say, further, the old house I told you that we lived in for fourteen years while we were building up the fertility of this soil, we sold for $10, after we got through with it. It is now a horse barn on the farm of our next neighbor and has been covered over.

"Eleven years from the time we started we paid the last $500 of our debt, all dug out of that farm, not $25 from any other source.

Thirteen years from the time we started, we carried off the first prize of $50 offered by the State Board of Agriculture of Ohio, for the best detailed report of the best and most profitably managed small farm in the state,—only thirteen years from the time we started on that rundown land, and no fertility brought from the outside; without any money; and meanwhile we had to live.

"Now I had arranged with the tenant the first year, before we went on there, to seed down a certain field. It had been under the plow for some time. I wanted it seeded so I could have some land to mow and he seeded half of it. It was only a little lot, about five acres. He seeded half with timothy and left the other half. That was his way of doing things, anyway. When we moved onto the farm later I naturally wanted to finish that seeding and get that field in some sort of shape for mowing. I went to my next neighbor, who lives there yet, and asked him what I had better use. I didn't know anything, practitically, about farming, and he advised me to try some clover seed. He said: 'So far as I know, none was ever sown on that farm. They have sowed timothy everlastingly, everybody, because it is cheap. I knew timothy wouldn't grow there to amount to anything If I were in your place I would try some clover.'

"I got the land prepared and sowed that clover alone, so as to give it a chance. I did have sense enough to mow off the weeds when they got six or eight or ten inches high perhaps, so that the clover could have a little better chance to grow. It happened to be a very wet season. I remember that distinctly. This was a lot near to the barn. I suppose what little manure they had hauled out had been mostly put on this land. With these favoring conditions the result was fairly good. Of course not half what we got later, but we got quite a little clover and when I came to mow it, and to mow that timothy at the other end, I could see I could draw the rake two or three times as far in the timothy as in the clover. There was more clover on an acre. A load of timothy would go in and a load of clover. When I fed it to the cows in winter I noticed when feeding clover for a number of days they gave more milk. I didn't know why. I don't know as anybody knew why then. There wasn't an experiment station in the land. We were following our own notions. But the cows gave more milk; I could see that plainly.

"A little later I had an experiment forced on me by accident. I tell you just how it came about. It resulted in putting a good many thousands in our pockets and I hope millions in the pockets of the farmers of America. Later I wanted to plant corn on this field, and, as I wanted to grow just as good corn as I could, I got out what manure we saved and put it on the land preparing for plowing. I knew there wouldn't be more than half enough to go over the field. I said to myself, if there was any good corn, I would like it next to the road where people would see it. Wouldn't any of you do it? I didn't have a dollar to hire any help. I paid one dollar that year for help, and it was awful hard to get that dollar. I began spreading that manure next to the road. The back half of the field was nearly out of sight. When I got half way back there wasn't any manure left and the back half didn't get any. Now it so happened that the timothy was on the front end of the field, and it got the manure. The clover on the back half didn't get any. It came about in the simple way I told you of. Naturally I didn't expect much corn where I hadn't put any manure, but what was my surprise to find it was just about as good on that clover end of the field without any dressing as on the timothy end with what I had been able to put on. It is only right I should say there wasn't much of the manure It was poor in quality because we couldn't get grain for the cows when we couldn't get enough for ourselves to eat. There wasn't much manure and it was pretty poor, but such as it was that was the result. More hay to the acre, better hay, increased fertility, some way, by growing this clover!

"Now let us go back a little. I think it was the second spring after we moved onto the place that I happened to be crossing the farm of my next neighbor, Mr. Holcombe, now dead. I found him plowing. He had been around a piece of land, I should judge five acres, half a dozen times. He was sitting on the plow, tired out,—too old to work anyway. He said, 'I wish you would take this land and put in some crop on the shares; I want to get rid of the work; I can't do it, and would like to let you have it in some way. All I want is that it should be left so I can seed it down in the fall again.'

"It was an old piece of sod he had mowed in the old eastern way until it wouldn't grow anything any longer. I don't suppose he got a quarter of a ton of hay to the acre. He wanted it plowed so he could re-seed it. I didn't know the value of the land, but, foolishly perhaps, as most people thought, offered him five dollars an acre for the use of it. I hadn't enough to do at home. I didn't have my land in shape so I could do much. We were working along as fast as we could. I thought I could do well if I had this job, and could perhaps make something off it. He agreed to it.

"I went home and got my team and plow, and finished the plowing. I remember making those furrows narrow and turning the ground well, a little deeper than it had been plowed before. I didn't realize what I was doing, then. I simply had been brought up to do my work well. I thought I was doing a good job, that was all. When I was through plowing I got my old harrow, a spike-tooth, and harrowed the ground. I had a roller. They were manufactured in our town. The firm bursted and I had a chance to buy one very cheap. I had a roller, harrow, and plow. That was all the tillage implements. The harrow had moved the lumps around a little. I ran the roller over the lumps; then harrowed, rolled, and harrowed. When the harrow would not take hold, I put a plank across and rode on it. I worked that land alternately until I had the surface as fine and nice as I could make it, two or three inches deep. The harrow would not take hold any longer and I had to quit. By and by a rain came. I didn't know anything about how to till land,—this spring fallow business—but I happened to hit it right. After it rained, I said that harrow will take hold better now. I loaded the harrow and got on it, and tore that ground up three or four inches deep.

"The harrow teeth were sharp. I harrowed and rolled it and my neighbor said, 'Terry, you are ruining that land, it will never grow anything any more, it will all blow away.' I reminded him of his bargain; I should raise what I pleased and take the crop home. Every little while, I can't remember how often, I would go over and harrow and roll that land. I probably plowed it the first week in April. For two months that was a sort of savings bank for my work. I would run over and work that land, occasionally, until, about the first week in June, I had it prepared just as mellow and fine and nice as it was possible to make it. It was nice enough for flower seeds."

"I builded better then than I knew. I had no idea what the result was going to be. When it was all ready, I sowed Hungarian grass seed. I wish you could have seen the crop. It grew four and a half or five feet high, as thick as it could stand on the land. I believe if I had thrown my straw hat, it would have staid on the top. It was enormous for that land. I had four big loads to the acre. You know what you can put on a load of Hungarian. When I went by the owner's house with those loads and took them to our barn, he was out there and he looked awfully sour. That man, to my knowledge, had never grown half as much to the acre since I had known of his being on the land, probably never more than one-third as much. Old run-out timothy sod; no manure, no fertilizer, nothing but the work,—this spring fallowing. I enjoyed the matter more, because he had told some of the neighbors he had got the start of that town fellow; I would never see five dollars an acre back, out of the land. That was his opinion of what I could raise.

"Hay was hay that fall, after a dry season. We live in a dairy section. The cows were there and had to be fed. I got $18 a ton for that hay in our barn, something like $70 per acre. I think the laugh was on the other side. That was my first awakening, along this line of tillage. Didn't know how it came about, didn't know anything about the fertility locked up in the soil, just the plain facts. I did so and so, and got such and such results. The next year Charlie Harlow, still living there, said, 'I wish you would put in some Hungarian for me this spring.' I said, 'What part of the crop?—I should want two-thirds.' He said he had an offer for half. I said, 'Then let him have it.' He replied, 'One-third of what you will raise is more than half of what he will raise.' He saw what I did on his brother-in-law's farm.

"The following year I had a piece of land ready to grow corn, I had cleared out the stumps and done the best I could to get it in shape. I plowed it just as soon as the ground was dry enough, about the first of April, that is. I worked it every little while just as nearly as I could as the Hungarian land had been worked, I harrowed and rolled, let it rest a while, then harrowed and rolled. I kept it up until my next door neighbor, Mr. Croy, had planted his corn, and it was four inches high and growing pretty well. Ours wasn't planted. A neighbor came and said, 'I am sorry for you, Terry, you don't know what you are about. You are fooling away your time. Your corn ought to have been in before this.' I was harrowing and rolling. I was determined to see whether I could do it over again. Some of the neighbors said it couldn't be done again.

"The fourth or fifth of June—too late, ordinarily, to plant corn with us—I put in the crop. I wish you could have seen it grow! It came up and grew from the word 'Go.' In four weeks it was ahead of any corn about. It went ahead of my neighbor's corn that was three or four inches high when ours was planted. We had a crop that, the farm in the condition that it was, was considered as something remarkable. They couldn't account for it, neither could I. All I knew was I had been working the ground so and so and getting such and such results.

"Let us go back once more. The first year that I moved onto that farm, the first fall, we had nine cows, and I wanted to save all of the manure. Now, there wasn't an experimental station in the land. I didn't know anything about the potassium or nitrogen in the liquid manure, but I had seen where it dropped on the land and how the grass grew. I thought it was plant food, and our land was hungry. I said, I must try and save this manure, and not have it wasted. I hadn't a dollar. What did I do? There was an old stable there that would hold ten cows. It was in terrible shape. It had a plank floor that was all broken. I tore it out. I hauled some blue clay. I filled the stable four or five inches deep with the blue clay, wet it, pounded it down, shaped it off and got it level, fixed it up around the sides, saucer shape, so it would hold water. Then I laid down some old boards (I couldn't buy new ones), and put in a lot of straw there and put my cows in. I saved all that manure the first year, all that liquid. I had twice as much, probably more, from the same number of cows as had been saved on that farm before, and it was much more valuable. That was the beginning the first winter, when I hadn't anything.

"For the horse stable I went to town and found some old billboards. It was new lumber, but had been used for billboards. After the circus the owner offered to sell the boards cheap, and to trust me. He was a carpenter, and he jointed them. We put them crosswise on the old plank floor, and when they got wet they swelled and became practically water tight. I even crawled under and saw that there was no liquid manure dropping down there. I drew sawdust and used for bedding. I saved the liquid of the horse stable. I didn't know it was worth three times as much, pound for pound, as the solid. I didn't know it was worth two times as much in the cow stable, pound for pound, as the solid. I found it out by experience.

"Now, when I was in town, before going on this farm, I worked for S. Straight & Son, the then great cheese and butter kings of the Western Reserve. I was getting over a thousand dollars a year in their office. They didn't want me to leave at all, but my wife and I took a notion to be independent, to work for ourselves, and we bought this old farm. We had a chance to work for ourselves, all right. The first year we worked from early in the morning until nine or ten o'clock at night, and then we tumbled into bed, too tired to think, to get up and do it over again. I worked in the field, taking out stumps and doing something, as long as I could see, and then helped my wife to milk. We would get our supper along about nine or ten o'clock. At the end of the year we had not one single dollar, after paying our interest and taxes,—not one dollar to show for our work. Do you wonder we were pretty discouraged?

"I met Mr. Straight one day. He said: 'Terry, things are not going very well in the office since you left. I wish you would come back. You are not doing much over on that farm that I can see. You are having a hard time. I will gladly give you $1,200 a year if you will come back into our office.' It was a great temptation. Think what it meant. To move back to town and have $100 a month. But I said, 'No, Mr. Straight; I can't do it.' I don't deserve any credit for it, friends: but I wasn't built that way. I can't back out. When I undertake anything I have got to go through. I would have been willing enough to leave that farm, if I had made a success of it, after I made a success of it, as I thought then; but I wasn't willing to give up, whipped—to acknowledge that I had undertaken that job and had to back out and go back to town to make a living.

"Some little incident sometimes will change the whole character of a man's life. I remember, when we were in very hard conditions, we were sitting under an apple tree in our door yard one evening. It is there yet. Two men from town went by. One of them said to the other, 'What is Terry going to do?' The other said, 'If Terry sticks to it he will make something out of that old farm.' Just as quick as a flash, friends, I said, 'Terry will stick to it.'

"You see what condition we were in. I began to put all these matters together. I had been taught how to. In college I had been trained to study and think, of course,—not to work with my hands. When I got onto the work at first I worked myself almost to death with my hands, and had no time to think or study; but gradually old methods came around again and I began to think and study. I said: 'Here, more hay to the acre, better hay, increased fertility by growing that clover, increased fertility by working that soil so much.' I didn't know why, but there was the fact. 'Now, isn't it possible to put these matters together and so work them out as to build up the fertility of this farm and make it blossom like the rose?'

"I began to work it out. What was the first step? I sold eight or nine cows to get a little money to start, thus cutting off practically our whole source of income. There was no other way I could get any money. We had to do some draining. A part of the land we could not do anything with until it was tile-drained. It took money to buy tile. I had to have a little help about the digging, although I like to boast that I laid every tile on my farm with my own hands. I buried every one and know it will stay there. They were all sound and hard and good. In all these years not one has ever failed, not one drain or tile. I worked day after day, in the rain, wet to the skin, because it had to be done. It was the foundation of our success.

"As I was coming here yesterday, and passed so much of your flat land, in need of drainage, I thought, drainage is the foundation of success for lots of these people, down here in southern Illinois. You can't do much until you have the water out of the land. Then you have a chance to do something with tillage and manure-saving and clover. But you throw away your efforts when you try to do this work on land that is in need of drainage.

"As fast as possible we fixed up this land. Of course, it took years. We hadn't money, and there were many things that had to be done,—changing fields, getting out stumps, doing drainage,—it all took time. I had my plans made and was working as fast as I could.

"Two things I did, to keep life in our bodies until we got ready to make some money. One was to cut off every bit of timber on the farm. Our neighbors laughed at us and prophesied rain and all that. There were two things in my mind. We had to have money to live on, and I managed to get quite a little of it in that way. In the next place we didn't have much of a farm, and I wanted the land for tillage. We can buy wood of the neighbors to-day, cheaper than we sold ours, so we never lost anything.

"Another way we got some money, as we went along, that helped us, was raising forage crops. I did not attempt to put in crops that required much hand labor. I raised Hungarian, and everything I could to be fed to cows. In our dairying section, with feed often scarce in the fall, farmers often had more stock than they could winter. We could pick up cows cheaply on credit and hold them. I could winter them for people, and the manure we used as a top dressing, to make the clover grow. Starting with a little piece of land, we spread out more and more, and got more and more enriched, and more and more growing clover, and by and by we got all the cultivated land growing it. Then we were ready for business.

"I am afraid to tell you Illinois farmers, with your great big farms, how large our farm was. We bought one hundred and twenty-five acres. We sold off all but fifty-five. That didn't help us, for the man who bought it was so poor he didn't pay us for over thirty years. Then the land went up in price and he was able to sell it for a good price and we got our money. Fifty-five acres were selected, the best we could for our purpose. Twenty acres were so situated as to have no value. Thirty-five acres were fairly good, tillable land, the best we could pick out. I began a system of rotation, after we got the land ready for it, of clover, potatoes, and wheat. My idea was to have the clover gather fertility to grow potatoes and wheat. I was going to make use of the tillage to help out all I could, and sold the potatoes and wheat, and then had clover again, and so on around the circle. Everybody said, of course I would fail. I didn't know but I would. It was the only chance and I had to take it.

"Of course it took quite a while to get this thing going. The first three or four years didn't amount to much. After six or eight years we were surprised at the result. We were getting more than we hoped for. In a dozen years the whole country was surprised. I remember when a reporter was sent from Albany, New York, to see what we were doing, and reported in the "Country Gentlemen." We had visitors by the score from various states, it made such a stir. They couldn't believe it was possible for a man to take land as poor as that, and make it produce so well. We had some they could see that had not been touched. As I told you, in eleven years we were out of debt. After about ten or eleven years we were laying up a thousand dollars a year, above all living and running expenses, from this land, raising potatoes and wheat. It doesn't seem possible to you, large farmers, but you can't get around the facts. In 1883 we laid up $1,700 from the land. But this was a little extra.

"We wanted to build a new house. We had lived in the old shell long enough. We had the money to pay cash down for the new house and to pay for the furniture that went into it. We paid $3,500 cash down, that fall, for the house and furniture, and every dollar taken out of the land. Only two or three years before that we paid the last of our debt. I had not done any talking or writing to speak of, at that time. I did not begin until 1882 I never went to an institute, and never wrote an article for a paper, except when called upon to do it. I never sought such a job and prefer to stay at home on my farm. It was only because I was called to do this work that I got into it. For twenty-one years I was never at home one week during the winter season. Farmers called for me and I didn't feel that I could refuse to go.

"Now, how did we do it? I told some of the things. Let us go down to the science of the matter little, now. I didn't know anything about the science at the time. That came later. Practice came first. We know now—of course, you all know—that clover has the ability, through the little nodules that grow on the roots, to take the free nitrogen out of the air to grow itself. You know about four-fifths of the air you are breathing is nitrogen in the form of gas, and clover has the ability to feed on that and make use of it. The other plants have not. I might illustrate it in this way: You can't eat grass; at least, you wouldn't do very well on it. But the steer eats grass and you eat the steer, so you get the grass, don't you? Your corn, wheat, oats, timothy, potatoes, so far as we know, can't touch free nitrogen in the air, but clover can and then feed it to those other crops.

"Let us look into how we got the phosphorus. On land that would not grow over six to eight bushels of wheat per acre we have succeeded once in growing forty-seven and three-fourths bushels to the acre, on all the land sowed, of wheat that sold away above the market price and weighed sixty-four pounds to the measured bushel, and never put on a pound of phosphorus. We got it from that tillage we told you about. Our land in northeastern Ohio is not very good naturally. It is nothing like what you have in this state. Most of you know that is the poorest land we have in the state in general, but we have a fair share of clay and sand in ours. That has helped us wonderfully. We have clay enough so that with our tillage we can make so far all the plant food available we want.

"Now, a little more about the tillage. I told you how we worked the surface of that ground and made it fine and nice. After five or six years, perhaps, of this kind of work, I got to thinking if I had some tool that would stir that ground to the bottom of the plowed furrow and mix it very deeply and thoroughly, I might get still better results out of the tillage. I happened to be in town one morning in the fall, when we had some wheat land (clover sod) plowed and prepared for wheat. I had harrowed and rolled it and made it as nice as I could.—It was what the neighbors would call all ready for sowing and more than ready. In town I saw a man trying to sell a two-horse cultivator. I think it was made in this State. It was the first one I ever saw—you can judge how long ago. It was a big, heavy, cumbersome thing,—a horse-killer. I thought, if I only had that, I knew I could increase the fertility of our soil still more. I hadn't any money. We hadn't got far enough that there was a dollar to spare. What did I do? I gave my note for $50 and took that cultivator home with me. I could have bought it for $35 in money, but I didn't have it. My wife didn't say a word when I got home. I have heard since that she did a lot of crying to think I would go in debt $50 more, and all for that thing.

"I got home about eleven o'clock and you can well suspect that I couldn't eat any dinner that day. I hitched up and went right to work, and told my wife I couldn't stop for any dinner. I rode that cultivator that day and tore up that field in a way land was never torn up in our section before. There was nothing to do it with. The soil would roll up and tumble over. After going lengthwise I went crosswise. A thousand hogs couldn't have made it rougher. The neighbors looked on and said that 'Terry would do 'most anything if you would only let him ride.' The worst of it was, I really didn't know but what they were right, and all he would get out of it was the riding. It was a serious thing. I had to wait until the harvest time before I could know.

"What was the result? I got ten bushels of wheat more per acre than had ever grown on the land before, without any manure or fertilizer having been applied since it grew the previous crop in the rotation. Clover had been grown. It was a clover sod. I didn't know how much came from the clover and how much from the tillage. I didn't care, they went together to get that result. I asked some of the old settlers how much had been grown there per acre during their recollection. They said twenty-three bushels was the most they had known. I got thirty-three. The neighbors said, 'It happened so, you can't do it again.' You know how they talk, to make out nothing can be done with an old farm. I was interested in doing it again. I paid that note and had a large margin of profit left, you see, out of the extra wheat. It all came right.

"The next year I took the next field in rotation and worked it in the same way, probably more. I got thirteen bushels more wheat per acre than ever grew before. Thirty-six bushels of wheat! Such a thing was never heard of in our section before; land that would not grow anything a dozen years ago. Do you wonder I have been an enthusiast on tillage since then? Why, they call me a crank sometimes. It is a good crank, as it has turned out prosperity for us.

"After a time I began to think, can't we carry this matter a little further? People generally don't cultivate their crops more than two or three times in a season. Can I cultivate more to advantage? I began to try it, six or eight times, eight or ten. I think there have been dry years when I have cultivated our potatoes as many as fifteen times. I don't believe we ever went through them when it didn't pay.

"I remember one fall, when it was a wet season. When the tops began to die and got to the point where I could see the space between the rows, I started the cultivators again. I had money then to hire men and I hired plenty of them. I started to cultivate between the rows. People said, ' What is the idiot doing now?' I said, 'He is going to raise five bushels more by doing that work, that it what he is after.'

"Now, remember, more hay to the acre, better hay, increased fertility by growing clover, increased fertility by working this land over and over in the different ways I have told you of. They used to send for me to talk on this subject, before I knew anything about it, except that I had done it. In Wisconsin, some twenty years ago, I helped at the first institute held in the state. They sent for me to come up. I told them what I was doing and how I thought it came about, what I thought clover was doing for me. When I was through I asked Professor Henry, who was in the audience, to tell me, honestly, what he thought about my talk. He said, 'As a farmer I believe you are right, but as a scientific man I dare not say so in public.'

"Professor Roberts came to my place one time, to investigate a little. I knew what he came for. I showed him around, and showed him the land we had not touched, not to this day. He was a surprised man. I remember the second crop of clover was at its best. It was above his knees. He says, 'This will make two tons of hay to the acre, and it is the second crop.' He didn't say but very little. I couldn't get him to talk much. He went home and began that system of experiments at Ithaca that has practically revolutionized the agriculture of the east—experiments in tillage. Pretty soon we had his book on the fertility of the soil. I think he got his inspiration from what he saw. He said to himself, seems to me, 'Terry has something that scientific men do not know.' He got samples of soil all over the state. They analyzed the soil and found what the average soil of New York contained. They found about four thousand five hundred pounds of nitrogen, six thousand three hundred pounds of phosphoric acid, and twenty-four thousand pounds of potash in an average acre eight inches deep; and they had been buying potash largely. (Laughter.)

"The farm we moved onto was the old Sanford homestead. Old Mr. Sanford lived there and brought up a large family. I think five of them boys. Every one of these boys left the farm just as soon as they could get away. There wasn't anything in farming for them. After we had been at work a dozen years or more and got things going nicely, they came back (one of them lives in Connecticut) and visited the old homestead. I remember Lorenzo said, 'It seems like a miracle. I don't know how you did it. We worked from daylight to dark, from one year's end to another, and never had anything. We boys used to be promised a holiday on the Fourth of July if the corn was all hoed. That was all we got. How on earth have you done these things?'

"Friends, there were three farms we bought. Old Mr. Sanford didn't know anything about but one. There was the air and the soil and there was the subsoil. He had been working only the soil, plowing it three or four inches deep, scratching it over, taking what came, and every year less and less came. The land had run down until the surface had quit producing. We took the same soil, put in clover and took the fertility out of the upper farm, the air, and out of the lower one, the subsoil, and put it into the second one. We plowed the surface soil a little deeper and deeper until we got it eight or nine inches deep instead of four. We worked it more and more, setting more and more of the available plant food in the soil free. That is how we did it.

"I say 'we' advisedly, because, friends, if I hadn't had a wife fully able and willing to do her part, and more, I would not have this story to tell."

CHAPTER XXXVIII

AN AWAKENING DREAM

"THE chores are all done," said Mrs. Johnston, as Percy began to take down his heavy work-coat about nine o'clock that evening.

"You ought not to have done them," he chided as he slipped his arm around her and drew her to the sofa.

"Tell me about the Institute," she said, stroking the hair from his forehead.

He told her of the professors who were there from the University and briefly reported the addresses he had heard.

"And I verily believe," he added, "that if Terry were to wake up some morning and find himself located on the "Barrens" of the Highland Rim of Tennessee, he would start out with the firm conviction that all he would need to do to become a successful farmer there would be to sow clover and then 'work the land for all that's in it.' But, after all, it is not so strange, perhaps, that one who has himself discovered and then utilized the power of clover and tillage to restore and increase the productive power of land rich in limestone, phosphorus and all other essential mineral plant food, should jump to the fixed and final conclusion that the same system of treatment is all that is needed to make any and all land productive. The fact that Terry's land (if equal to the nearby New York land) contained two thousand three hundred pounds of phosphorus in the plowed soil of an acre when he began to work it out, while the soil of the Tennessee "Barrens" contains only about one hundred pounds, does not disturb him or modify his opinion so long as his personal experience is limited to his own land.

"Terry's problem was easier than Mr. West's on his Virginia farm, where the soil is acid and hence limestone must be used liberally in order that clover and other legumes may be grown successfully. Even the supply of phosphorus and other mineral elements is probably greater in Terry's farm in northeastern Ohio than in the soil of Westover.

"Our problem is even more difficult, because we must not only increase the supply of active organic matter, although we have a reserve of old humus far above that contained in the Terry or West farms; but in addition we need more limestone than Mr. West and then we must add the phosphorus. Of course the surface washing is a serious factor on Westover, but perhaps our tight clay subsoil is worse.

"But I learned at least two things that I shall try to profit by. One of these was from Governor Hoard's lecture on 'Cows Versus Cows, and the man behind the cow'; and the other is that we must do more work on the land."

"Oh, Percy, I am so sorry you went. How can you possibly do more work than you have been doing?"

"I may need to hire more," he replied; "and, of course, that will further increase our expenses, but, it will surely pay to do well what we try to do."

"When does my boy expect to get married?" she asked, softly, as she gently stroked his hair.

"I am married," he replied.

She looked at him in wonder.

"Mother mine, I thought that you knew I was married."

"Your face is blank sincerity, as usual," she said smiling, "but you never deceive me with your voice. Your voice reveals every attempt at deception. Tell me what you mean."

His voice was sincere now. "I am married to a farm and laboring together with God. After hearing Terry's talk, I am more than ever determined to continue to do my part, working in the light as He gives me the power to see the light."

"Percy, dear," she asked, "did you know the bride whose wedding cards you received yesterday?"

"Don't you remember what I told you of Adelaide West, Mr. West's daughter?" he queried.

"I thought so," said the mother. She stepped to Percy's home-made desk, and from one of the pigeon holes, drew out a bunch of letters, and selected the top and bottom letters from the pile.

"Here are the first and last letters you have received from Mr. West. Did you ever see this?" She drew out a crumpled piece of paper and placed it in his hand.

"*Her Grandma had not consented,*" he read. "What does that mean?"

"I do not know and I did not know when I read it three years ago. It came in your first letter from Mr. West. I thought you had not found it in the envelope, but you gave me the letter to read and I found it. I left it in the letter, but never till to-day did I feel that I ought to mention it to you. Yesterday you received a letter with two cards; but you read only one of them to me."

"But I saw the other was only the wedding announcement, and I left them both in the letter for you to read."

"And I read them both," she said. "Read this."

Percy took the card and slowly read:

_Mr. and Mrs. Clarance Voit

Announce the marriage of their daughter

Ameila Louise

to

Professor Paul Strongworth Barstow_

She watched his face but saw no sign. She kissed his forehead and then pointed to the writing, "*With Grandma's Compliments,*" saying, "I do not know what this means, but I thought my boy might be getting too careless, when he fails to read even the wedding announcement of college professors, sent to him by such a good friend as Grandma West may intend to be."

Percy looked into his mother's face as if to read her thoughts.

"I think I understand what you have in mind," he said. "Mr. West has mentioned once or twice that Adelaide was teaching school, but I supposed that she was trying to earn enough to buy her own wedding outfit."

"Perhaps that is true," replied the mother, "and perhaps she is already married or soon to be married; but I thought you ought to know that she had not married Professor Barstow, lest you might allude to it in your letters to Mr. West."

CHAPTER XXXIX

HONEY WITHOUT WAX

"WELL, I reckon the cowboy's gone back to 'tend to his cows," remarked the grandmother to Adelaide, as she returned from taking Percy to Blue Mound and found the old lady sitting on the lawn bench apparently enjoying the mild late November weather. "Did you leave him at the station or see him off?"

"Neither," Adelaide replied, sitting down beside her. "The train was late, and he insisted on coming back with me to the first turn, and then stood and watched till I came within sight of home at the next turn. I doubt if he is back to the station yet."

"He reminds me, Pet, of the Latin definition you gave for _sincere," _remarked the grandmother. "Pure honey without wax, wasn't it?"

"Oh, no, Grandma. Not pure honey. It says nothing about honey. Sine is the Latin for _without, _and _cera _means _wax; _so that our word _sincere, _taken literally from the Latin, means *without wax.*"

"Oh, yes, I see now; but let me tell you, Adelaide, I think that professor of yours is right smart wax."

"Why, Grandma! I never heard you say such a thing. You know papa and mamma like Professor Barstow and I think I like him too, and,—and he has papa's consent, and mamma's consent."

"Well, you never heard me say such a thing before and you won't ever hear it again, but he hasn't got my consent. I think he's some wax, but I reckon you think he's some honey, and I know he thinks he's some punk'ns. Of course, your father would like an English or Scottish nobleman for a son-in-law, or at least a college professor with a string of ancestry reaching across the water; but the Henry's prefer to make their own reputations as they go along, and I doubt if Patrick ever saw England or Scotland. I tell you, Adelaide, a pound of gumption will make a better husband than a shipload of ancestry, and I just hope you will more than like your husband, that's all."

With that the old lady arose and walked to the house.

CHAPTER XL

INSPIRATION

WESTOVER,

March 14, 1907.

Mr. Percy Johnston,

Heart-of-Egypt, Ill.

MY DEAR Friend:—We were delighted to receive your interesting letter of March 2, describing the Farmer's Institute. I have been to two such meetings in Virginia, but they are devoted to fruit and truck and dairying, and no one seems to know much about our soils. I appreciate more and more every year the absolute knowledge you helped me to secure concerning Westover, where we had been working in the dark for two centuries. I am sure you will succeed on Poorland Farm,—just as confident as any one can be in advance of actual achievement; and I expect to see the time when Richland Farm will be a more appropriate name.

I only wish you could see my alfalfa. I have been seeding more every year and now have sixty acres. It has come through winter in fine condition and it will be a fine sight by Easter. Here's a standing invitation to take Easter dinner with us, or any other dinner, for that matter, if you ever come East.

I am planning to sow about forty acres more alfalfa this year. A writer for the _Breeder's Gazette _visited us last summer, and he said some of our alfalfa was as good as any he had ever seen in California. He said ground limestone was plainly what we need for alfalfa at Westover, but he thought some phosphorus would also help on the less rolling areas, where the alfalfa is not so good as where you found more phosphorus.

Lime and raw rock phosphate make the difference between clover and no clover.

I can get ground limestone for $2.90 a ton now, delivered at Blue Mound in bulk in carload lots. We are hoping to get it still lower, and I think we will, for some of the big lime manufacturers, such as the company at Riverton, are making plans to furnish ground limestone; and the railroad companies are likely to make better rates, or the State will do so for them.

It is truly a lamentable situation, when our hills and mountains are full of all sorts of limestone, and our exhausted lands are crying for that more than anything else. We understand, even better than you, that everybody is poor in a country where the land is poor; and it should be to the greatest interest of the railroad companies as well as to all other industries, to unite in an effort to make it possible for every landowner to apply large amounts of limestone to his land,—the more the better,—and no one should expect any large profit from the business; but wait till the benefit is produced on the land,—wait till the farmer has his increased crops, and some money from the sale of those crops. Then the railroads can make profit hauling those crops to market and hauling back the necessary supplies, and even the luxuries, which the farmer's money will enable him to buy and pay for. Then the factory wheels will turn; for, as you told us, the Secretary of Agriculture reports that eighty-six per cent. of all the manufactured products are made from agricultural raw materials.

There is no danger but what the railroads and manufacturers and commercial people will get their share out of the produce from the farms; but it is absolutely sure that, when the farms fail to produce, then there is no profit for any of them, and the last man to starve out will be the farmer himself, for he can live on what he raises even though he has nothing left to sell.

We are all well. My son Charles is still bookkeeping for a Richmond firm, but he is becoming greatly interested in my alfalfa, and says he sometimes wishes he had taken an agricultural course instead of the literary at college. His grandmother says she reckons the agricultural college could give him about all the literature he needs keeping books for a hides and tallow wholesale company; and I am coming to believe that she is about right. I still remember that the dative of indirect object is used with most Latin verbs compounded with _ad, ante, con, in, inter, ob, post, pre, pro, sub, _and _super, _and sometimes _circum; _but it would have been just as easy for me to have learned forty years ago that the essential elements of plant food are carbon, oxygen, and hydrogen; nitrogen, phosphorus, and potassium; magnesium, calcium, iron and sulfur; and possibly chlorin; and I am sure that the culture of Greek roots and a knowledge of Latin compounds have been of less value to me during the forty years than the culture of alfalfa roots and even a meager knowledge of plant-food compounds have been during the last three years.

Adelaide is teaching; Frank is in the academy; and the younger children are all in school.

We shall always be glad to hear from you.

Very respectfully yours,

CHARLES WEST.

"That is an exceptionally good letter," said Mrs. Johnson, as Percy finished reading.

"Not for Mr. West," he replied. "His letters are always good, always helpful and encouraging, almost an inspiration to me. Mr. West is in many ways a very exceptional man. If he had not been tied down all his life to a so-called worn-out farm of a thousand acres, he might just as well have been the Governor of the State. Even in spite of himself he has been practically forced to accept some very responsible public offices, but the financial sacrifice was too great to permit his retaining them very long. I never realized until I was nearly through college that the trustees of our own University devoted a large amount of time to that public service with no financial remuneration whatever. They are merely reimbursed for their actual and necessary travelling expenses."

"Well, if I were a young man about your age, this letter would be an inspiration to me," said his mother.

"You mean his suggestion about changing the name of our farm?"

"No, I mean his possible suggestion about changing the name of his daughter."

Percy was silent.

"How can I tell anything from your blank face? Why do you not speak?"

"You will have to show me," said Percy.

"Will you accept his invitation?"

"Oh, Mr. West always closes his letters with an invitation for me to visit them if I ever come East. There is nothing exceptional or unusual in that."

"The letter is very exceptional," she repeated, "insomuch that if there is no understanding there is no misunderstanding, and if there is some misunderstanding there was no intention. When Mrs. Barton says: 'Do come over when you can,' there is no invitation intended and no acceptance expected; but when Mrs. McKnight says: 'Can't you and your son come over and take supper with us Thursday evening,'—well that is an invitation to come. In the case of Mr. West's letter, perhaps you had an invitation to spend the Easter vacation at Westover when his daughter will be at home,—and perhaps not."

Percy was silent and his mother quietly waited.

"In any case," he said, "I cannot afford to go this spring. We never were so short of funds. I almost begrudged the railroad fare I paid to go to the Institute."

"I have agreed to agree with you regarding the matter of hiring more help on the farm if you need it," she said; "for it is easily possible to lose by saving. There are some things which should never be influenced by financial considerations. It is more than three years since your Eastern trip. You need a rest and a change. It would be entirely commonplace for you to spend the Easter time in Virginia. You ought to see the country in the spring; and you ought especially to be interested in Mr. West's sixty acres of alfalfa. Expectations are always followed either by realization or by disappointment, either of which my noble son can bear."

Her fingers passed through his hair as she kissed his forehead.

"The only question is, whether you would enjoy a visit to Westover," she continued. "You have insisted that the Winterbine deposit remain in my name, but I have written and signed a check against that reserve for $100, and you have only to fill in the date and draw the amount at the County Seat whenever you wish. If you go, express my regards to the ladies, and especially remember me to the grandmother."

CHAPTER XLI

THE KINDERGARTEN

HEART-OF-EGYPT, ILLINOIS,

November 9, 1909.

Hon. James J. Hill,

Great Northern Railroad Company, St. Paul, Minnesota.

MY DEAR SIR:—I have read with very great interest your article in the November _World's Work _on "What We Must Do to be Fed." I wonder if you read *The American Farm Review!* In the editorial columns of that journal, issue of October 28, 1909, occurs the following:

"The pessimist always assumes that every man who quits farming for some other business does so because there is something the matter with the farm. Mr. James J. Hill has recently considered the question and decided that, unless the farmer and his family can be confined on the land and be compelled to do better work than they have been doing, the balance of the population must starve to death. The bug-aboo of impending decadence raised by such talk is based upon a wrong assumption, inadequate statistics, and a failure to comprehend the evolutional movement in agriculture."

The evolutional movement means, of course, that we are different from other people. Have not England, Germany and France run their lands down until they produce only fourteen bushels of wheat per acre and have we not steadily built ours up to an average yield of thirty bushels? Other peoples wear out their soil because they fail to have part in the evolutional movement; whereas, did we not come to America and at once begin to make our rich land richer than it ever was in the virgin state? Do you not know, Sir, that the oldest lands in America are now the richest, most productive, and most valuable? We admit, of course, that the Bureau of Soils of the United States Department of Agriculture reports the common level upland loam soil of St. Mary country, Maryland, to be valued at $1 to $3 an acre, and the same kind of land in Prince George county, adjoining the District of Columbia, to be worth $1.50 to $5; but do you not know the American evolutional movement could easily move all those decimal points two places and at once make those values read from $100 to $500 an acre. And likewise, it would be a very simple matter to change the yield of corn in Georgia from eleven bushels per acre and have it read one hundred and ten bushels. Why not, if an acre of corn in the adjoining State of South Carolina has produced two hundred and thirty-nine bushels in one season? Do you not see that this simple evolution would also put plate glass in the thousands of windowless homes now inhabited by human beings, both white and colored, in the state of Georgia?

There is another phase of this evolutional movement which should not be overlooked. There is already fast developing in this country a class of people who can live and grow fat on hot air, and they will tell you that your only trouble is poor digestion, and they are glad that they can see the bright side of things and enjoy life in this glorious country, assured that the future will take care of itself. Have not all other great agricultural countries rapidly gotten into this evolutional movement until all their people live on Easy Street?

I have a letter from a missionary in China, a former schoolmate, Clarence Robertson, who resigned the position of Assistant Professor of mechanical engineering in Purdue University in order to accept in the largest sense the Master's specific invitation to "Go ye, therefore, and teach all nations."

This letter was written in February, 1907 and contained the following statement regarding the famine district in which the writer was located:

"At the present time the only practical thing to do is to let four hundred thousand people starve, and try to get seed grain for the remainder to plant their spring crops."

I think we have failed utterly, Mr. Hill, to lay special emphasis upon either the evolutional or the emotional in agriculture. Is it not probable that a superabundance of emotion would even permit the constitution to wave the bread requirement in the bread-and-water-with-love diet? As a cure for pessimism the emotional tonic is strongly recommended.

On the other hand, there are some people who are even too emotional, people who are inclined to sit up and take notice when the mathematics and statistics are spread out in clear light and plainly reveal the fact that the time is near at hand when their children may lack for bread. (They already lack for meat and milk and eggs in many places). To ally any feeling of this sort that might tend to excite those who are so emotional as even to love their own grandchildren, some sort of soothing syrup should be administered. A preparation put out by the Chief of the United States Bureau of Soils and fully endorsed by the great optimist, the Secretary of Agriculture, is recommended as an article very much superior to Mrs. Winslow's. As a moderate dose for an adult, read the following extracts from pages 66, 78, and 80 of Bureau of Soils Bulletin 55 (1909), by the Chief of the Bureau:

"The soil is the one indestructible, immutable asset that the nation possesses. It is the one resource that cannot be exhausted; that cannot be used up."

"From the modern conception of the nature and purpose of the soil it is evident that it cannot wear out, that so far as the mineral food is concerned it will continue automatically to supply adequate quantities of the mineral plant foods for crops."

"As we see it now, the main cause of infertile soils or the deterioration of soils is the improper sanitary conditions originally present in the soil or arising from our injudicious culture and rotation of crops. It is, of course, exceedingly difficult to work out the principles which govern the proper rotation for any particular soil."

"As a national asset the soil is safe as a means of feeding mankind for untold ages to come. So far as our investigations show, the soil will not be exhausted of any one or all of its mineral plant food constituents. If the coal and iron give out, as it is predicted they will before long, the soil can be depended on to furnish food, light, heat, and habitation not only for the present population but for an enormously larger population than the world has at present."

"Personally, I take a most hopeful view of the situation as respects the soil resources of our country and of the world at large. I cannot bring myself to believe that the discouraging reports that have been issued from time to time as to the threatened deterioration of our soils, as to the exhaustion of any particular element of fertility, will ever be realized."

Sweeten to taste, and repeat the dose if necessary.

If you desire mathematical proof that we can always continue to take definite and measurable amounts of plant food away from the limited supplies still remaining in our American soils and still have enough left to supply the needs of all future crops, let it be understood:

That y = x

Then xy = X3

And xy-y2 = x3-y2

Or y(x-y)=(x + y) (x-y)

Hence, y = x + y

Thus, y = 2y

Therefore, 1=2

Now cube both sides of the last equation and:

1=8

Multiply by one hundred and sixty, the number of pounds of phosphorus still remaining in the common upland soil of Southern Maryland, and behold:

160 =1280

Thus the soil again becomes the equal of the $200 corn belt land,—Q. E. D.

Fortunately, Mr. Hill, you have not found it "exceedingly difficult to work out the principles which govern the proper rotation" that "actually enriches the land."

Seriously, I hope you will permit me to take this opportunity to say that I deplore, as must all right-minded and clear-thinking men, the occasional petty criticisms which attribute to you some selfish motive for the honest and noble stand you have taken concerning the importance of immediate action and of a widespread, far-reaching, and generally effective movement looking toward, not the conservation, but the restoration, and permanent preservation of American soils. According to the Scriptures, there is a sin which God, Himself, will not forgive; namely, the sin of imputing bad motives to the one who does right from motives only good and pure.

Thoughts that deserve a place of honor in American history you have expressed in the following words:

"The farm is the basis of all industry, but for many years this country has made the mistake of unduly assisting manufacture, commerce, and other activities that center in cities, at the expense of the farm. The result is a neglected system of agriculture and the decline of the farming interest. But all these other activities are founded upon the agricultural growth of the nation and must continue to depend upon it. Every manufacturer, every merchant, every business man, and every good citizen is deeply interested in maintaining the growth and development of our agricultural resources. Herein lies the true secret of our anxious interest in agricultural methods; because, in the long run, they mean life or death to future millions; who are no strangers or invaders, but our own children's children, and who will pass judgment upon us according to what we have made of the world in which their lot is to be cast."

True and noble thoughts are these, from the master mind of a great statesman; for there are statesmen who neither grace nor disgrace the Halls of Congress.

Your article contains twenty-eight pages of wholesome reading matter and instructive illustrations, and, in addition, about one page, I regret to say, of misinformation that will do much to destroy your otherwise valuable contribution to agricultural literature.

Briefly you have shown very clearly and very correctly that the present practice of agriculture in America tends toward land ruin, and that, with our rapidly increasing population, with continued depletion of our vast areas of cultivated soils, and with no possibility of any large extension of well-watered arable lands, we are already facing the serious problem of providing sufficient food for our own people.

You summarize your conclusions along this line in the following words:

"We have to provide for a contingency not distant from us by nearly a generation, but already present. The food condition presses upon us now. The shortage has begun. Witness the great fall in wheat exports and the rise of prices. Obviously it is time to quit speculating about what may occur even twenty or thirty years hence, and begin to take thought for the morrow. As far as our food supply is concerned, right now the lean years have begun."

It is certain that the time is near when our food supplies shall become inadequate if our present practices continue, but the enforced reduction in animal products will at least postpone the time of actual famine in America. I keep in mind always that we are feeding much grain to domestic animals, an extremely wasteful practice so far as economy of human food is concerned; because, as an average, animals return in meat and milk not more than onefifth as much food value as they destroy in the corresponding grain consumed; and, as we gradually reduce the amounts of grain that are fed to cattle, sheep, and swine, we shall also gradually increase our human food supply. Ultimately our milk-producing and meat-producing animals will be fed only the grass grown upon the non-arable lands and possibly some refuse forage not suitable for human food or more valuable for green manure, unless we modify our present practice and tendency, which we can do if the proper influences are exerted by the intelligent people of this country, and thus make possible the continuation of high standards of living for all our people.

I keep in mind, too, that much of the food taken into the average American kitchen is wasted, and that progress in the science of feeding the man will ultimately prevent this waste and, by adding to this better preparation and combination of foods, will increase to some extent the nutritive value of our present food supply.

The serious fact remains, however, that our older lands are decreasing in productive power and, in spite of what may be accomplished by such methods of conservation, we are now facing a rapidly approaching shortage of food supplies for the rapidly increasing population of these United States; and you have put me and all other American citizens under lasting obligations to you for your frankness, good sense, and true patriotism in thus pointing out n advance our great national weakness.

According to the statistics of the United States Government, a comparison of the last five years reported in this century with the last five years of the old century, shows, by these two five-year averages, that our annual production of wheat has increased from about five hundred million to seven hundred million bushels: that our annual production of corn has increased from two and one-quarter billion to two and three-quarter billion bushels; that our wheat exports have decreased from thirty-seven per cent. to seventeen per cent. of our total production; that our corn exports have decreased from nine per cent. to three per cent. of our total production; and yet the average price of wheat, by the five-year periods, has increased thirty-one per cent., and the average price of corn has increased ninety-one per cent., during the same period.

The latest Year Book of the Department of Agriculture (1908) furnishes the average yields of wheat and corn for four successive ten-year periods, from 1866 to 1905. By combining these into two twenty-year periods this record of forty years shows that the average yield of wheat for the United States increased one bushel per acre, while the average yield of corn decreased one and one-half bushels per acre, according to these two twenty-year averages.

If we consider only the statistics for the North-Central states, extending from Ohio to Kansas and from "Egypt" to Canada, the same forty-year record shows the average yield of wheat to have increased one-half bushel per acre, while the average yield of corn decreased two bushels per acre.

Thus, notwithstanding the great areas of rich virgin soils brought under cultivation in the West and Northwest during the last forty years, notwithstanding the abandonment of great areas of wornout lands in the East and Southeast during the same years, notwithstanding the enormous extension of dredge ditching and tile drainage, and, notwithstanding the marked improvement in seed and in the implements of cultivation, the average yield per acre of the two great grain crops of the United States has not even been maintained, the decrease in corn being greater than the increase in wheat, and not only for the entire United States, but also for the great new states of the corn belt and wheat belt.

(Seasonal variations are so great that shorter periods than twenty-year averages cannot be considered trustworthy for yield per acre.)

Meanwhile, the total population of the United States increased from thirty-eight millions in 1870 to seventy-six millions in 1900, or an increase of one hundred per cent. in thirty years; and the only means by which we have been able to feed this increase in population has been by increasing our acreage of cultivated crops and by decreasing our exportation of foodstuffs; and I need not remind you that the limit to our relief is near in both of these directions. But have we decreased our exportation of phosphate? Oh, no. On the contrary, under the soothing influence of the most pleasing and acceptable doctrine that our soil is an indestructible, immutable asset, which cannot be depleted, our exportation of rock phosphate has increased during the years of the present century from six hundred and ninety thousand tons in 1900, to one-million three hundred and thirty thousand tons in 1908, an increase of practically one hundred per cent., in accordance with the published reports of the United States Geological Survey.

But I am writing to you, Mr. Hill, not only to thank you for what you have said and shown in the twenty-eight pages above referred to, but also in part to repay my obligation to you by giving you some correct information, which I am altogether confident you will appreciate; namely, that, while you are a graduate student or past master in your knowledge of the supply and demand of the world's markets, you are just entering the kindergarten class in the study of soil fertility, as witness the following extracts from the one erroneous page of your article.

"Right methods of farming, without which no agricultural country such as this can hope to remain prosperous, or even to escape eventual poverty, are not complicated and are within the reach of the most modest means. They include a study of soils and seeds, so as to adapt the one to the other; a diversification of industry, including the cultivation of different crops and the raising of live stock; a careful rotation of crops, so that the land will not be worn out by successive years of single cropping; intelligent fertilizing by the system of rotation, by cultivating leguminous plants, and, above all, by the economy and use of every particle of fertilizing material from stock barns and yards; a careful selection of grain used for seed; and, first of all perhaps in importance, the substitution of the small farm, thoroughly tilled, for the large farm, with its weeds, its neglected corners, its abused soil, and its thin product. This will make room for the new population whose added product will help to restore our place as an exporter of foodstuffs. Let us set these simple principles of the new method out again in order:

_"First—_The farmer must cultivate no more land than he can till thoroughly. With less labor he will get more results. Official statistics show that the net profit from one crop of twenty bushels of wheat to the acre is as great as that from two of sixteen, after original cost of production has been paid.

_"Second—_There must be rotation of crops. Ten years of single cropping will pretty nearly wear out any but the richest soil. A proper three or fiveyear rotation of crops actually enriches the land.

_"Third—_There must be soil renovation by fertilizing; and the best fertilizer is that provided by nature herself—barnyard manure. Every farmer can and should keep some cattle, sheep, and hogs on his place. The farmer and his land cannot prosper until stock raising becomes an inseparable part of agriculture. Of all forage fed to live stock at least one-third in cash value remains on the land in the form of manure that soon restores worn-out soil to fertility and keeps good land from deteriorating. By this system the farm may be made and kept a source of perpetual wealth."

Your _first principle _will be agreed to and emphasized by all; but it should be kept in mind that the large farms are frequently better tilled than the small farms. The $200 land in the corn belt is usually "worked for all that's in it." It is tile-drained and well cultivated, and the best of seed is used. If more thorough tillage would increase the profits, these corn-belt farmers would certainly practice it.

It ought to be known (1) that as an average of six years the Illinois Experiment Station produced seventy and three-tenths bushels of corn per acre with the ordinary four cultivation, and only seventy-two and eight-tenths bushels with additional cultivation even up to eight times; and (2) that the average yield of corn in India on irrigated land varies from seven bushels in poor years to twelve bushels in good seasons, and this is where the average farm is about three acres in size.

One Illinois farmer with a four-horse team raises more corn than ten Georgia farmers with a mule a piece on the same total acreage Fertile soil and competent labor are the great essentials in crop production. A mere increase in country population does not increase the productive power of the soil.

The farms down here in "Egypt" average much smaller than those in the corn belt of Illinois, but our "Egyptian" farms are nevertheless poorly tilled as a rule and some of them are already becoming abandoned for agricultural purposes.

Certainly the land should always be well tilled, but tillage makes the soil poorer, not richer. Tillage liberates plant food but adds none. "A little farm well tilled" is all right if well manured, but it should not be forgotten that the men who consider "Ten Acres Enough" are market gardeners, or truck farmers, who are not satisfied until in the course of six or eight years they have applied to their land about two hundred tons of manure per acre, all made from crops grown on other lands.

All the manure produced in all the states would provide only thirty tons per acre for the farm lands of Illinois. In round numbers there are eighty million cattle and horses in the United States, and our annual corn crop is harvested from one hundred million acres. All the manure produced by all domestic animals would barely fertilize the corn lands with ten tons per acre if none whatever were lost or wasted; and, if all farm animals were figured on the basis of cattle, there is only one head for each ten acres of farm land in the United States.

Your _second principle _is, that "a proper three or five-year rotation of crops actually enriches the land."

I hope the God of truth and a long-suffering, misguided people will forgive you for that false teaching. If there is any one practice the value of which is fully understood by the farmers and landowners in the Eastern states and in all old agricultural countries, it is the practice of crop rotation. Indeed, the rotation of crops is much more common and much better understood and much more fully appreciated in the East than it is in the corn belt. Practically all we know of crop rotation we have learned from the East. Every old depleted agricultural country has worn out the soil by good systems of crop rotation. I once took a legal option of an "abandoned" farm in Maryland (beautiful location, two miles from a railroad station, gently undulating upland loam, at $10 per acre) that had been worn out under a four-year rotation of corn, wheat, meadow and pasture. A few acres of tobacco were usually grown in one corner of the corn field, and clover and timothy were regularly used for meadow and pasture. Wheat, tobacco and livestock were sold, and manure was applied for tobacco and so far as possible for corn also. In the later years of the system the ordinary commercial fertilizer was also applied for the wheat at the usual rate of two hundred pounds per acre, this having become a "necessity" toward the end of this slow but sure system of land ruin.

The "simple principles" of your "new method" were understood and practiced in Roman agriculture two thousand years ago; and they included not only thorough tillage, careful seed selection, regular crop rotation, and the use of farm manure, but also the use of green manures. Thus Cato wrote:

"Take care to have your wheat weeded twice—with the hoe, and also by hand."

And again Cato wrote:

"Wherein does a good system of agriculture consist? In the first place, in thorough plowing; in the second place, in thorough plowing; and, in the third place, in manuring."

Varro, who lived at the same time as Cato, wrote as follows:

"The land must rest every second year, or be sown with lighter kinds of seeds, which prove less exhausting to the soil. A field is not sown entirely for the crop which is to be obtained the same year, but partly for the effect to be produced in the following; because there are many plants which, when cut down and left on the land, improve the soil. Thus lupines, for instance, are plowed into a poor soil in lieu of manure. Horse manure is about the best suited for meadow land, and so in general is that of beasts of burden fed on barley; for manure made from this cereal makes the grass grow luxuriantly."

Virgil wrote in his *Georgics:*

"Still will the seeds, tho chosen with toilsome pains, Degenerate, if man's industrious hand Cull not each year the largest and the best."

It was in 1859 that Baron von Liebig wrote as follows, regarding these and similar _ancient _teachings:

"All these rules had, as history tells us, only a temporary effect; they hastened the decay of Roman agriculture; and the farmer ultimately found that he had exhausted all his expedients to keep his fields fruitful and reap remunerative crops from them. Even in Columella's time, the produce of the land was only fourfold. It is not the land itself that constitutes the farmer's wealth, but it is in the constituents of the soil, which serve for the nutrition of plants, that this wealth truly consists."

Suppose, Mr. Hill, that a successful American farmer should tell you that your bank account will actually increase if you will give from three to five members of your family the privilege of writing checks instead of following the single checking system. "But," you will ask, "doesn't rotation produce a larger aggregate yield of crops than the single crop system?" Certainly, and, likewise, a rotation of the check book will produce a larger aggregate of the checks written; but the ultimate effect on the bank deposit is the same as on the natural deposit of plant food in the soil, and finally the checks will not be honored. Indeed, it would be a fine sort of perpetual motion if we could actually enrich the soil by the simple rotation of crops, and thus make something out of nothing.

Consider, for example, the common three-year rotation, corn, wheat, and clover. A fifty-bushel crop of corn removes twelve pounds of phosphorus from the soil; the twenty-five bushel wheat crop draws out eight pounds; and then the two-ton crop of clover withdraws ten pounds, making thirty pounds required for this simple rotation. The most common type of land in St. Mary county,

Maryland, after two hundred years of farming, contains phosphorus enough in the soil for five rotations of this simple sort. Mathematically that is all the further traffic in rotations that soil can bear. Agriculturally that soil has refused to bear any sort of traffic, whether single or in rotations, and has been abandoned for farm use except where fertilized.

These crops would remove from the soil one hundred and twenty-four pounds of nitrogen in the corn and wheat, and the roots and stubble of the clover would contain forty pounds of nitrogen. Now, if the soil furnishes seventy-six pounds of nitrogen to the corn crop and forty-eight pounds to the wheat crop, will it furnish forty pounds to the clover crop, or as much as remains in the roots and stubble? If so, how does the rotation actually enrich the soil in nitrogen?

You will be interested to know that there are many exact records of the effect upon the soil of the rotation of crops. This particular three-year rotation has been followed at the Ohio Agricultural Experiment Station for thirteen years, and the average yield of wheat has been, not twenty bushels, not sixteen bushels, but eleven bushels per acre, where no plant food was applied; although where farm manure was used the wheat yielded twenty bushels, and with manure and fine-ground natural rock phosphate added the average yield of wheat for the thirteen years has been more than twenty-six bushels per acre. The corresponding yields for corn are thirty-two, fifty-three and sixty-one bushels, and for clover they are one and two-tenths, one and six-tenths and two and two-tenths tons of hay per acre.

You will wish to know also that the Ohio Station has conducted a five-year rotation of corn, oats, wheat, clover, and timothy for the last fifteen years, both with and without the application of commercial plant food. As an average of the fifteen years the unfertilized and fertilized tracts have produced, respectively:

30 and 48 bushels of corn

32 and 50 bushels of oats and 27 bushels of wheat .9 and 1.6 tons of clover

1.3 and 1.8 tons of timothy

In 1908 the unfertilized land produced nine-tenths ton of clover, while land treated with farm manure produced three and two-tenths tons per acre.

You will welcome the information that the average yield of wheat on an Illinois experiment field down here in "Egypt," in a four-year rotation, including both cowpeas and clover, has been eleven and one-half bushels on unfertilized land, fourteen bushels where legume crops have been plowed under, and twenty-seven bushels where limestone and phosphorus have been added with the legume crops turned under; and that the aggregate value of the four crops, corn, oats, wheat, and clover, from another "Egyptian" farm, has been $25.97 per acre on unfertilized land, and $54.24 where limestone and phosphorus have been applied.

In your very busy and very successful railroad experience, you may have overlooked the reports of the Pennsylvania Agricultural Experiment Station, showing the results of a four-year rotation of crops that has been conducted with very great care for more than a quarter of a century. These, you will agree, are exactly such absolute data as we sorely need just now when facing the stupendous problem of changing from an agricultural system whose equal has never been known for rapidity of soil exhaustion to a system which shall actually enrich the land. By averaging the results from the first twelve years and also those from the second twelve years, in this rotation of corn, oats, wheat, and hay (clover and timothy), we find that the yields have decreased as follows:

Corn decreased 34 per cent.

Oats decreased 31 per cent.

Wheat decreased 4 per cent.

Hay decreased 29 per cent.

Appalling, is it not? It is the best information America affords in answer to the question, Will the rotation of crops actually enrich the land?

No, Sir. We cannot make crops nor bank accounts out of nothing. The rotation of crops does not enrich the soil, does not even maintain the fertility of the soil. On the contrary, the rotation of crops, like the rotation of your check book, actually depletes the soil more rapidly than the single system; and, if you ever have your choice between two farms of equal original fertility, one of which has been cropped with wheat only, and the other with a good three or five-year rotation, for fifty years, take my advice and choose the "worn-out" wheat farm. Then adopt a good system of cropping with a moderate use of clover, and you will soon discover that your land is not worn out, but "almos' new lan" as a good Swede friend of mine reported who made a similar choice. But beware of the land that has been truly worn out under a good rotation, which avoids the insects and diseases of the single crop system, and also furnishes regularly a moderate amount of clover roots which decay very rapidly and thus stimulate the decomposition of the old humus and the liberation of mineral plant food from the soil.

Perhaps you have heard of Rothamsted. If not, your kindergarten teacher is at fault. A four-year rotation of crops has been followed on Agdell field for more than sixty years. An average of the crop yields of the last twenty years reveals:

(1) That the yield of turnips has decreased from ten tons to one-half ton per acre since 1848.

(2) That the yield of barley has decreased from forty-six bushels to fourteen bushels since 1849.

(3) That the yield of clover has decreased from two and eight-tenth tons to one-half ton since 1850.

(4) That the yield of wheat has decreased from thirty bushels to twenty-four bushels since 1851, wheat, grown once in four years, being the only crop worth raising as an average of the last twenty years.

No, Sir. Neither optimism, nor ignorance, nor bigotry, nor deception can controvert these facts.

Do you know that the people of India rotate their crops? They do; and they use many legumes; and some of their soils now contain only a trace of phosphorus, too small to be determined in figures by the chemist. Do you know there are more of our own Aryan Race hungry in India than live in the United States?

Do you know that Russia regularly practices a three-year rotation and actually harvests only two crops in three years, with one year of green manuring? Yes, and the average yield of wheat for twenty years is only eight and one-quarter bushels per acre.

Think on these things.

Your _third principle _is, that "of all forage fed to live stock at least one-third in cash value remains on the land in the form of manure that soon restores worn-out soil to fertility and keeps good land from deteriorating. By this system the farm may be made and kept a source of perpetual wealth."

I grieve with you; pity 'tis, 'tis not true.

No, Sir. Neither crops nor animals can be made out of nothing, and no independent system of livestock farming can add to the soil a pound of any element of plant food, aside from nitrogen, and even this addition is due to the legume crops grown and not to the live stock.

Under the best system of live-stock farming about three-fourths of the nitrogen, three-fourths of the phosphorus, and one-third of the organic matter contained in the food consumed can be returned to the land if the total excrements, both solid and liquid, are saved without loss. Of course, the produce used for bedding can all be returned, but it could also be returned without live stock.

Under a good system of crop rotation with all grain sold from the farm it is possible to return to the soil more than one-third of the phosphorus and more than one-half of the organic matter contained in the crops, and even as much nitrogen as all of the crops remove from the land in the grain sold. Thus, with a four-year rotation of wheat, corn, oats, and clover, and a catch crop of clover grown with the wheat and turned under late the following spring for corn, we may plow under three tons of clover containing one hundred and twenty pounds of nitrogen, in return for the one hundred and nineteen pounds removed from the soil for the twenty-five bushels of wheat, fifty bushels of corn, and fifty bushels of oats. These amounts of grain and the two bushels of clover seed might be sold from the farm, while the two and one-half tons of straw, one and one-half tons of stalks, and three tons of clover might be returned to the land. These amounts aggregate seven tons of organic matter, or the equivalent of seventeen tons of manure, measured by the nitrogen content, or of twenty-four tons, measured by the content of organic matter. To replace the twenty-two pounds of phosphorus sold from the farm in the grain of these four crops would require the expenditure of sixty-six cents at the present prices for raw phosphate delivered at Heart-of-Egypt.

I have no doubt you will be glad to have your attention called to the fact that the world does not live wholly, or even largely, upon meat and milk. Bread is the staff of life, and I note from your _World's Work _article that you prefer to have the bread made of wheat. Thus, most farmers must raise and sell grain and vegetables.

If no independent system of live-stock farming can add a pound of phosphorus to the one hundred and sixty pounds still remaining in the great body of the level uplands constituting forty-one per cent. of St. Mary county, and forty-five thousand seven hundred and seventy acres of Prince George county, Maryland, adjoining the District of Columbia, nor even maintain the phosphorus supply in our good lands, then what must we do to be fed?

Manifestly, we should make large use of legume crops for the production of farm manure or green manure; and, manifestly, America should stop selling every year for five million dollars enough raw phosphate for the production of more than a billion dollars' worth of wheat. How long can we afford to give away a thousand millions for five millions?

Our annual corn crop is nearly three billion bushels, while the estimated value of all the timber on the still remaining federal lands is only one billion dollars. Again, our three trillion tons of coal is sufficient for an annual consumption of half a billion tons for six thousands years, whereas the United States Geological Survey has estimated that at the present rate of increase in mining and exportation our total supply of high-grade phosphate will be exhausted in fifty years. It seems to me that about ninety per cent. of the talk about conservation of natural resources is directed toward ten per cent. of the resources, when we remember the soil as the foundation of all agriculture and all industry.

Do you know, Mr. Hill, that, at the Second Conservation Conference called by the President of the United States, Doctor Van Hise, of the University of Wisconsin, was the only man to raise his voice in the interests of the common soils of America? For three days the statesman and experts discussed the forests, forests, forests, and the waters, waters, and the coal and iron; and for fifteen minutes President Van Hise pleaded for the conservation of phosphate, _the master key to all our material prosperity; _and he was called a crank with a hobby.

With deep respect, I am,

Very sincerely yours,

PERCY JOHNSTON

CHAPTER XLII

ADVANCE INFORMATION

HEART-OF-EGYPT, November 14, 1909.

DEAR father and mother: I can scarcely realize that I have been an "Egyptian" for almost two years. I feel that the time has been shorter than two months of school-teaching.

Percy is so encouraged with the crops that I rejoice with him, although I could never weep with him unless I weep for joy. He says the crops needed only that I should stroll over the fields with him; that they would grow rapidly if I only looked at them. Think of it—I drove the mower to cut hay,—not all of the 80 acres, to be sure, but I cut where it yielded two tons per acre. That is on No. 4, where Percy applied his first cars of limestone. I wish you could have seen the untreated strips—no clover and only half a ton of weedy timothy, while the rest of No. 4 and No. 6 were clean hay of mixed alsike and timothy. Percy says that No. 4 produced as much real hay last year as all the rest of the farm has produced since he came, and that the hay crop this year is worth as much for feed as all that has been harvested during the previous five years; and the cattle and horses seem to agree with him.

We sold our main lot of hogs for $654, and have another lot to go later. We are getting so many horses and cattle on the place, that we are going out of the hog business.

Percy says that hogs belong more properly in the corn belt, than in the wheat and fruit belt. You know the year I came the corn crop was on No. I, which had never grown anything but corn, oats, and wheat, so far as we can learn; and the corn was so poor the hogs ate most of it in two months' time. During the same two months the price of hogs dropped from 7 to 4-1/2 cents, so that the hogs were worth no more after eating the corn crop than they were before.

Next year we are to have corn on No. 4, and Percy says it will be the first time that corn has had a "ghost of a show to make a decent crop" since he bought the place. The spring before we were married he reseeded that forty, sowing mixed alsike and timothy. The clover came on finely, evidently because the scanty growth of clover the year before had at least allowed the field to become thoroughly infected with the clover bacteria. There was no clover on the unlimed strip. So we say that limestone and bacteria brought clover. The hay and other feed has made manure enough so that No. 4 has been completely covered with six tons per acre, and the phosphate has also been applied; so with manure and phosphate on clover ground we hope to grow corn next year, if we have good weather.

The phosphate has also been put on some of the other forties. I convinced him that the money will pay a higher rate of interest in phosphate than it would in the savings bank, even if he put it on before manure and clover could be plowed under. The experiments of several states show this very conclusively.

The corn is on No. 3 this year and it is the best crop in the six years. Percy says the "Terry Act" (which means lots of work in preparing the land) is some help, but he thinks the phosphate shows against the check strips. The young wheat on No. 2 is looking fine, and with both limestone and phosphate on that field and the extra work on the seed bed, we hope for a better crop than we have ever grown on a full forty; even though we must depend solely upon our reserve stock of nitrogen for the crop. We are all about as jealous of that reserve stock of organic matter and nitrogen as we are of the Winterbine bank account.

I cannot forget how Percy tried to persuade me to postpone our wedding for a year because, as he said, the hogs had taken his corn crop and given nothing in return for it; and above all how he objected to my reimbursing the Winterbine reserve from my teacher's wages to the extent of $250, which he had drawn in part to tide over the hard times, and in part to come to see me that Easter. But I am glad to have him still insist upon it that that uncertain venture proved his best investment, even if he does tease by adding that it paid one hundred and fifty per cent. net profit at Winterbine.

We are selling some cows this fall,—trying to weed out our herd by the Babcock test which shows that "some cows don't pay their board and keep," to quote Governor Hoard's lecture on "Cows versus Cows," which Percy heard at Olney the winter Professor Barstow was married. The "versus cows" are worth only $45.

I cannot tell you how I have enjoyed the summer. Sir Charles Henry is the dearest child, and his grandmother insists upon it that it is better for me to help Percy in the field with such light work as I can do, and I am out for a few hours every day when the weather is good. Percy's mother is such a dear. I am sure she could be no more sweet and loving to an own daughter. She had Percy all to herself for so long that I was really afraid she might not like to share him with me, but Percy says that it was his mother who persuaded him to make us that Easter visit. We tell her that she hasn't much use for either of us now, and that we are likely to get jealous because Charles Henry gets so much of her affection.

I forgot to tell you of Percy's four-acre patch of wheat. He said it is so long to wait till 1912 for his first wheat crop on land that had grown clover at least once during historic times that he thought he would fix up a little patch to grow a crop of wheat, just to see how real wheat would look; or, as he sometimes says, to see how wheat grows in "Egypt" when it has a ghost of a chance.

He treated a four-acre patch down by the wood's pasture with limestone, phosphorus, and farm manure, did the "Terry Act" in preparing the seed bed, and drilled in a good variety of wheat, on October 17,—a little later than he likes to finish sowing wheat. It came up with a good stand but did not make very much fall growth, partly owing to the dry weather. In the spring the man came across the patch and reported to Percy that the wheat was mighty small and he guessed it was "gone up," although it seemed to be all alive. Percy said that he would not worry about it if it were alive because the wheat would find something to please it when it really woke up in the spring. I reckon it did, for a neighbor passed on his way to town in early May and called over the fence to Percy that his patch of rye down by the woods was looking fine. Well the four acres yielded 129-1/2 bushels, or a little more than thirty-two bushels per acre. Percy said if he could have eighty acres of it and sell it for $1.18 a bushel, the same as he got for the last he sold, it would amount to twice the original cost of the land—and then some.

Mr. Barton asked him if he could not raise "just as good crops with good old farm manure," and if he could not build up his whole farm with farm manure. Percy said yes, but he would need three thousand tons for the first application. Mr. Barton then suggested that that was more than the whole township produced.

No. 5 has been in pasture for three years, clover and grass having been seeded in 1906, even though the wet weather had prevented the seeding of wheat the fall before, and the ground was left too rough for the mower. Percy hopes to have that forty completely covered with manure by the time he will be ready to apply the phosphate and plow it under for the 19 I I corn crop.

Now your "Egyptian" son has just read over this long, long letter, and he says that if I were a real wise old farmer I would not begin to talk about results before a single forty acres of grain had had a ghost of a chance to make a crop. He says that every bushel of corn, oats, and wheat that this old farm has produced during the last six years has been wholly at the expense of the meager stock of reserve nitrogen still left in the soil after seventy-five years of almost continuous effort to "work the land for all that's in it" He says that we have no right to expect really good crops until after the second rotation is completed, because the clover grown during the first rotation does not have a fair show, the limestone not yet being well mixed with the soil, the phosphorus supply being inadequate, the inoculation or infection being imperfect, and no provision whatever having been made to supply decaying organic matter in advance of the first clover crop. I think he is right as usual and I promise to give no more advance information hereafter except upon inquiry, at least not until 1918, when the first wheat crop will be grown on land which has been twice in clover. We are mighty sorry not to be able to be with you for Thanksgiving or Christmas, but really we cannot go to the expense; our house is so small (we just must build a larger barn) and our home equipment is so meager that, in the words which you will remember Percy told us his mother credited to Mrs. Barton, I feel that as yet I must say,

"Do come over when you can."

Your happy, loving daughter,

ADELAIDE.

P.S.—Three big oil wells, belonging to the class called "gushers," have been struck about seven or eight miles from Poorland Farm. We are all getting interested except Percy. He says he does not want any oil wells on his six rotation forties or in the wood's pasture, but he might let them bore in the twelve-acre orchard, which has never produced but one crop that paid for itself, and the profit from that is about all gone for the later years of spraying.

The first oil boom in Illinois was at Casey where they struck oil six or eight years ago, but they say the wells there are dry already and they have to go back to farming again to get a living. Of course if we could get a hundred-barrel well on every ten acres and get a royalty of $400 a day for a few years, it would help out nicely, but the oil business is uncertain and short-lived, whereas, to quote Percy "the soil is the breast of Mother Earth, from which her children must always draw their nourishment."

Some have spoken to Percy about the coal right, but he says if there are ten thousand tons of coal per acre under Poorland Farm, he will save it for Charles Henry before he will allow anyone else to take it out for less than ten cents a ton. He says that just because the United States Government was generous enough to give the settler three hundred and twenty acres of land, and foolish enough to throw in with it three million tons of coal if it happened to lie beneath, is no reason why he should sell it to any coal company or coal trust at the rate of ten tons for one cent, which is the same as ten dollars per acre for the coal right. He says if Uncle Sam ever wants to assume his rightful ownership of all coal, phosphate deposits, or other minerals whose conservation and proper use is essential to the continued prosperity of all the people, then our coal shall be his; but, if he does not want it then he will consider nothing less than leasing on the basis of a royalty of ten cents a ton to be paid to him, his heirs, and assigns, etc.; but even then he wants enough coal left to hold up the earth, so that there will be no interference with the tile drains which he expects sometime to put down at an expense exceeding the original cost of the land. With much love,

ADELAIDE.

P.S.—Percy sends his love to grandma and a photograph for Papa, from which you will see that on such land as ours no limestone or phosphate means no clover.—A. W. J.

The author takes this occasion to say to the kind reader who has had the patience and the necessary interest in the stupendous problem now confronting the American people, of devising and adopting into general practice independence systems of farming that will restore, increase, and permanently maintain the productive power of American farm lands,—to those who have read thus far the _Story of the Soil _and who may have some desire for more specific and more complete or comprehensive information upon the subject,—to all such he takes this occasion to say that this volume is based scientifically upon "Soil Fertility and Permanent Agriculture."

This little book is intended as an introduction to the subject; the other may be classed as technical, but nevertheless can be understood by any one who gives it serious thought. This book tells the true story of the soil, for which the other gives a thousand proofs.

Grateful acknowledgment is here expressed that even the measure of success thus far attained on Poorland Farm has been possible largely through the co-operation of a beloved brother, Carl Edwin, the man who does a world of work, ably assisted by "Adelaide."